NEUROCASE
2008, 14 (1), 1–6

Neuroscience and crime

Hans J. Markowitsch

Physiological Psychology, University of Bielefeld, Bielefeld, Germany

Jurisprudence will profit considerably from methods and applications of the neurosciences. In fact, it is proposed that the neurosciences will provide unique possibilities and advantages in understanding motivations and causes for staying lawful or for becoming unlawful. Neuroscientific models on brain–behavior interactions have profited considerably from the advent of neuroimaging techniques and genetic analyses. Furthermore, advances in interdisciplinary investigations, which combine conventional psychological and sociological explorations with biological examinations, provide refined insights into the question 'What makes us tick?' (Weiskrantz, 1973, *British Journal of Psychology, 64*, 511–520). The search for such interactions from the time of the nineteenth century to the present is briefly surveyed and it is concluded that the interdisciplinary approaches within and across neuroscientific fields will lead and have already led to a considerable expansion of our knowledge. The articles in this issue devoted to highlighting the latest neuroscience research related to criminal behavior underline the power of this new approach.

Keywords: Functional brain imaging; Lie detection; Genetic analyses; Crime; Jurisprudence; Forensic psychiatry.

INTRODUCTION

Results of neuroscientific research influence many other scientific fields, which in former times neglected the existence of the brain and the interdependence between brain and behavior. The advent of functional neuroimaging methods was especially recognized as a new stage in unraveling the mechanisms of perceptions, motivations, emotions, and memories. The previous privacy of thoughts, ideas, and wishes was brought towards greater light by neuroimaging. It became possible to find correlates between psychogenic disturbances, such as dissociative amnesia (Markowitsch, 2003) and conversion paralysis (Burgmer et al., 2006), and changes in brain metabolism. Similarly, it was possible to relate the success of treatments to changes in brain metabolism (Markowitsch, Kessler, Weber-Luxenburger, Van der Ven, & Heiss, 2000).

The last decade has been characterized by recognizing neurosciences as central to various other scientific fields. Expressions like 'neuroeconomy', 'neuroethology', 'neuroanthropology', 'neurotheology' 'neuropsychoanalysis', 'neurolinguistics', etc., indicate that the input from the 'neuro-field' is highly welcomed by other disciplines. No wonder then that also criminology – though with hesitation – has recognized that it might profit from the progress in the neurosciences (Markowitsch & Kalbe, 2007; Zeki & Goodenough, 2006).

The law in general is conservative and relies on traditions. The uniform of British judges who have to appear with a medieval wig characterizes such traditions, as do the robes judges need in many other countries when appearing in court. And recent reports emphasize that judges rely – in spite of multiple counterevidence (e.g., Dodson & Krueger, 2006; Loftus, 2003) – more on the subjective reports of eyewitnesses than on objective data, based for example on measurements using scientific equipment (Busey & Loftus, 2006; Saks & Koehler, 2005). Only DNA analyses may be seen as an exception.

Address correspondence to Hans J. Markowitsch, Physiological Psychology, University of Bielefeld, P.O.B. 10 01 31, D-33501 Bielefeld, Germany (E-mail: hjmarkowitsch@uni-bielefeld.de).

DOI: 10.1080/13554790801994756

HISTORICAL AND PRESENT ATTEMPTS TO FIND RELATIONS BETWEEN DEVIANT BRAINS AND DEVIANT BEHAVIOR

In the neurosciences, on the one hand, there has been a long – though over time not very successful – tradition of trying to find abnormalities in the brains of criminals (reviewed in Markowitsch, 1992). Lenin, for example, was considered to have been an 'athlete of associations', because extraordinarily large pyramidal neurons were found in his cortical tissue (Vogt, 1929); and portions of Einstein's brain were repeatedly examined and special features were reported and found to be in line with his extraordinary mathematical-physical imaginative power (Anderson & Harvey, 1996; Colombo, Reisin, Miguel-Hidalgo, & Rajkowska, 2006; Diamond, Scheibel, Murphy, & Harvey, 1985; Witelson, Kigar, & Harvey, 1999). On the other hand, since the case of Phineas Gage was published in the nineteenth century (Bigelow, 1850; Harlow, 1848, 1869; see also Damasio, Grabowski, Frank, Galaburda, & Damasio, 1994), people became aware of the fact that a damaged brain may have lasting consequences on personality dimensions. Subsequently, one of the first female doctors, Leonore Welt from Switzerland, published in 1888, in the German journal *Deutsches Archiv für klinische Medicin*, an article entitled 'On character changes in human beings after lesions of the frontal lobe' (Ueber Charakterveränderungen des Menschen infolge von Läsionen des Stirnhirns). In this article she described how people, after frontal lobe damage, changed from being accurate and reliable persons to unstable, grumpy and unreliable subjects.

While these people usually did not become criminals, there are recent descriptions emphasizing a close relationship between frontal lobe damage and deviant behavior. Above all, the investigations of Adrian Raine point to both structural and metabolic changes in the frontal lobes of murderers (Raine, 2001; Raine, Lencz, Bihrle, LaCasse, & Colletti, 2000; Yang et al., 2005), as well as in other brain structures (Raine et al., 2003, 2004). Similarly, the brains of pedophilic offenders show structural deviations from normal in the right amygdala and diencephalus (Schiltz et al., 2007). Particular attention was devoted to the case of a father who suddenly became pedophilic and was imprisoned, and who later turned out to have a tumor in his right frontal lobe. Removal of the tumor resulted in reinstallation of his normal, non-pedophilic personality (Burns & Swerdlow, 2003).

Numerous related studies could be cited that all point to close relations between structural changes and functional consequences. So, the calcification of both amygdalae can result in a changed perception of threatening stimuli and can lead to less fearful behavior (Markowitsch et al., 1994; Siebert, Markowitsch, & Bartel, 2003). Furthermore, an overactive amygdala may lead to sudden aggressive attacks (Mark & Ervin, 1970).

These descriptions and searches for structure–function relations in psychopaths can be regarded as a 'new phrenology'. In fact, soon after the decay of classical phrenology, Cesare Lombroso, in 1876, published a very influential book in which he described physiognomic features, walking abnormalities (Figure 1), gestures, etc., all of which were in his eyes indicative for characterization of the 'born criminal'. Other scientists of that time were convinced that they had established the 'carnivorous type' of human brain (Benedikt, 1876) or had weighed brains in order to observe relationssships with deviant behavior (Meynert, 1867). However, the current methodological and technical progress and advantages, on the bases of which inferences are drawn, differ widely from those of the early periods of brain research.

Consequently, present-day research frequently emphasizes the existence of distinct and predictable relations between brain changes and tendencies for criminal behavior (Basserath, 2001; Blake, Pincus, & Buckner, 1995; Bufkins & Luttrell, 2005), as well as those between genes (e.g., Caspi et al., 2002) and hormones (Klötz, Garle, Granath, & Thiblin, 2006; Popma et al., 2007) and behavior. However, an even stronger focus is laid on the shaping and changing of neuronal networks by environmental influences. Such studies range from the influence of raising children on their behavior (Fries, Ziegler, Kurian, Jacoris, & Pollak, 2005; Murray & Farrington, 2005) to those measuring relations between degrees of experience and changes in volumes of brain structures (Biegler, McGregor, Krebs, & Healy, 2001; Maguire et al., 2000; Sheline, Gado, & Price, 1998; Sheline, Gado, & Kraemer, 2003; Vermetten, Schmahl, Lindner, Loewenstein, & Bremner, 2006; Winter & Irle, 2004). In line with early work from psychoanalysis (Freud, 1910; Spitz & Wolf, 1946), recent studies such as that of Fries and co-workers (2005) found that early experience determines later behavior to an extraordinary extent. Children, raised in extremely aberrant social environments (Eastern European orphanages) and later adopted by American

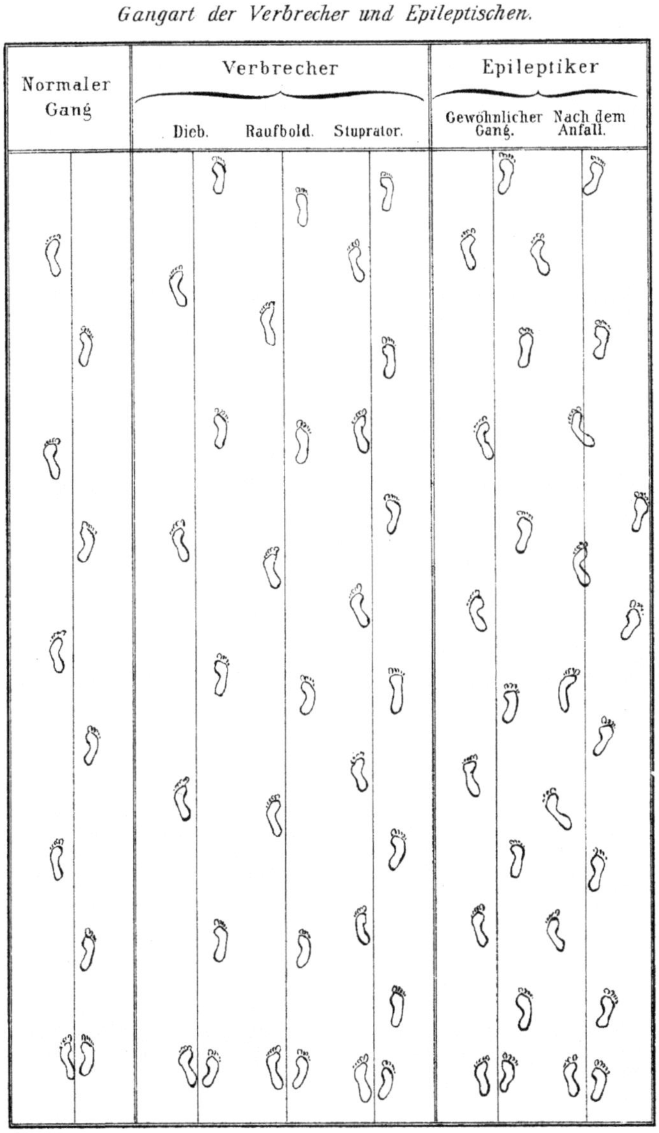

Gangart der Verbrecher und Epileptischen.

Figure 1. Example from Lombroso (1876, Gangart der Verbrecher und Epileptischen: Walking of criminals and epileptics), demonstrating differences between normal walking (Normaler Gang), the walking of criminals (Verbrecher: Dieb/thief, Raufbold/rowdy, Stuprator/rapist), and epileptics (Epileptiker: Gewöhnlicher Gang/ordinary walking, Nach dem Anfall/after the seizure).

parents, exhibited a lack of binding hormone releasers (oxytocin, vasopressin) even after years within their new, advantaged environment.

The impact of poor and disadvantageous environments on the likeliness of becoming criminal was the focus of much of the work of Dorothy Lewis and her co-workers (Lewis, 1998; Lewis et al., 1985; Lewis, Pincus, Feldman, Jackson, & Bard, 1986, Lewis, Pincus, Lovely, Spitzer, & Moy, 1987; Lewis, Yeager, Blake, Bard, & Strenziok, 2004). In this work, the social background of condemned criminals – usually death candidates – was studied and, together with neurological and psy-

chiatric features, related to their career as criminals. In one study the authors could demonstrate that, in all 15 prisoners waiting for execution, severe skull damage was found, together with further neurological signs in most of them. Six had schizoid psychoses, two were manic-depressive. In a similar study 18 other prisoners, also waiting to be executed, were studied. Again, all of them demonstrated multiple neurological and psychiatric damage or deviations. Six of those, who all were 18 years or younger, already had had birth complications. Seventeen of the 18 had had severe head traumata as children or juveniles. All manifested frontal lobe dysfunctions, 15 had psychiatric disturbances such as depression, schizo-affective or hypomanic disturbances, two were mentally retarded, one suffered from parasomnia and dissociation. All but one came from families who were extremely prone to violence and abuses, and in whom psychiatric diseases had been prevalent for generations.

Lewis et al. write that a primary aim of their studies is to emphasize ethical implications for the US justice system. Bioethicists and philosophers have emphasized that true autonomy – the ability to make reasonable choices – requires the capacity, not only to think about instinctive wishes and impulses, but as well, to control these and consequently do the right thing and avoid doing wrong things. Unfortunately, they continue to argue, not all brains have the same ability to foresee the consequences of behavior or of monitoring and controlling it. Lewis and co-workers then pose the question whether these juveniles could have performed virtuously, or whether the combination of psychopathology, frontal lobe dysfunction, and 'models' of violent peers, continuously acting around them, hindered them to think even once, of what they did at a given moment in time.

APPROACHING THE TRANSLUCENT HUMAN BEING

Insights from current research in the neurosciences therefore, in the first instance, provide strong evidence for determinism – the brain determines behavior and the environment changes the brain (cf., e.g., Sie & Wouters, 2008; Smilansky, 2000; Walter, 2002; Wegner, 2002). Secondly, such insights open many new windows for the law (cf. the discussion of Greene & Cohen, 2006). It becomes possible to state advantageous and disadvantageous conditions for human development and – in consequence – for the likeliness to stay within the rules of a given society or to deviate from it. Predictions of individual future behavior can be made much more firmly than in previous times and will be supported by so-called objective measures – above all functional neuroimaging. 'Lie detection' and 'mind reading' are just two of a number of fields, which, together with molecular biology (e.g., genetic analyses) open the road for what some people have named the translucent human being. The following articles provide examples of this new adventure of combining criminal and neuroscientific investigation. They also point to implications for successful prevention strategies which may become much more mandatory in the near future than they had been in the past.

The review of Dressing, Sartorius and Meyer-Lindenberg focuses on the possibilities of identifying offenders with antisocial personality disorders and sex offenders, stressing the impact of combined analyses, which aside from genetic analyses include other biological as well as psychological and social factors and variables. Anneliese Pontius then describes a single case with a 'Limbic Psychotic Trigger Reaction', a complex behavioral reaction, induced by sudden alterations in neuronal activity in specific brain regions, frequently leading to severe, but transient aggressive acts. Another patient with a similarly bizarre forensic pattern is reported by Kalbe and co-workers. This patient killed two of her children because she had the feeling that they lacked empathy, and she assumed that the same would hold true for their children. Functional brain imaging revealed, however, that she in principle was capable of feeling empathy.

Simone Reinders focuses on the topic of dissociative identity disorder, divided into three types of assumed origins: *iatrogenic,* originating from psychotherapeutic treatment; *traumagenic,* originating from severe childhood traumata; and *pseudogenic,* a simulated condition based on the intention to malinger. She discusses the possibilities of distinguishing between these origins on the basis of brain imaging data and to inform the judiciary accordingly.

Three papers, namely those of Kozel, Hakun et al., and Spence and Kaylor-Hughes, address the topic of applying functional brain imaging to lie detection. Bles and Haynes then discuss a very recent application of functional brain imaging, namely 'mind reading' – stating from the pattern of brain activations what a subject might perceive, think or do. And finally Strüber and Roth review, from an integrated point of view, the current status and possibilities of neuroscience and crime.

REFERENCES

Anderson, B., & Harvey, T. (1996). Alterations in cortical thickness and neuronal density in the frontal cortex of Albert Einstein. *Neuroscience Letter, 210,* 161–164.

Bassarath, L. (2001). Neuroimaging studies of antisocial behaviour. *Canadian Journal of Psychiatry, 46,* 728–732.

Benedikt, M. (1876). Der Raubthiertypus am menschlichen Gehirne [The carnivorous type of the human brain]. *Centralblatt für die medicinischen Wissenschaften, 42*, 930–933.

Benedikt, M. (1879). *Anatomische Studien an Verbrecher-Gehirnen* [Anatomical studies on the brains of criminals]. Wien: Wilhelm Braumüller.

Biegler, R., McGregor, A., Krebs, J. R., & Healy, S. D. (2001). A larger hippocampus is associated with longer-lasting spatial memory. *Proceedings of the National Academy of Sciences of the USA, 98*, 6941–6944.

Bigelow, H. J. (1850). Dr. Harlow's case of recovery from the passage of an iron bar through the head. *American Journal of the Medical Science, 39*, 13–22 (and 1 table).

Blake, P. Y., Pincus, J. H., & Buckner, C. (1995). Neurologic abnormalities in murderers. *Neurology, 45*, 1641–1647.

Bufkins, J. L., & Luttrell, V. R. (2005). Neuroimaging studies of aggressive and violent behavior: Current findings and implications for criminology and criminal justice. *Trauma Violence and Abuse, 6*, 176–191.

Burgmer, M., Konrad, C., Jansen, A., Kugel, H., Sommer, J., Heindel, W., Ringelstein, E. B., Heuft, G., & Knecht, S. (2006). Abnormal brain activation during movement observation in patients with conversion paralysis. *NeuroImage, 29*, 1336–1343.

Burns, J. M., & Swerdlow, R. H. (2003). Right orbitofrontal tumor with pedophilia symptom and constructional apraxia sign. *Archives of Neurology, 60*, 437–440.

Busey, T. A., & Loftus, G. (2006). Cognitive science and the law. *Trends in Cognitive Sciences, 11*, 111–117.

Caspi, A., McClay, J., Moffitt, T. E., Mill, J., Martin, J., Craig, I. W., et al. (2002). Role of genotype in the cycle of violence in maltreated children. *Science, 297*, 851–854.

Colombo, J. A., Reisin, H. D., Miguel-Hidalgo, J. J., & Rajkowska, G. (2006). Cerebral cortex astroglia and the brain of a genius: A propos of A. Einstein's. *Brain Research Reviews, 52*, 257–263.

Damasio, H., Grabowski, T., Frank, R., Galaburda, A. M., & Damasio, A. R. (1994). The return of Phineas Gage: Clues about the brain from the skull of a famous patient. *Science, 264*, 1102–1105.

Diamond, M. C., Scheibel, A. B., Murphy, G. M., & Harvey, T. (1985). On the brain of a scientist: Albert Einstein. *Experimental Neurology, 88*, 198–204.

Dodson, C. S., & Krueger, L. E. (2006). I misremember it well: Why older adults are unreliable witnesses. *Psychonomic Bulletin & Reviews, 13*, 770–775.

Freud, S. (1910). *Über Psychoanalyse. Fünf Vorlesungen gehalten zur 20jährigen Gründungsfeier der Clark University in Worcester Mass. September 1909* [On psychoanalysis. Five lectures given on the occasion of the celebration of the 20th anniversary of the foundation of Clark University in Worcester, MA. September 1909]. Leipzig: F. Deuticke.

Fries, A. B. W., Ziegler, T. E., Kurian, J. R., Jacoris, S., & Pollak, S. D. (2005). Early experience in humans is associated with changes in neuropeptides critical for regulating social behavior. *Proceedings of the National Academy of Sciences of the USA, 102*, 17237–17240.

Greene, J., & Cohen, J. (2006). For the law, neuroscience changes nothing and everything. In S. Zeki, & O. Goodenough (Eds), *Law and the brain* (pp. 207–226). Oxford: Oxford University Press.

Harlow, J. M. (1848). Passage of an iron rod through the head. *Boston Medical and Surgical Journal, 39*, 389–393.

Harlow, J. M. (1869). *Recovery from the passage of an iron bar through the head.* Boston, MA: D. Clapp and Son.

Klötz, F., Garle, M., Granath, F., & Thiblin, I. (2006). Criminality among individuals testing positive for the presence of anabolic androgenic steroids. *Archives of General Psychiatry, 63*, 1274–1279.

Lewis, D. O. (1998). *Guilty by reason of insanity. A psychiatrist explores the minds of killers.* New York: Balantine Publishing Group.

Lewis, D. O., Moy, E., Jackson, L. D., Aaronson, R., Restifo, N., Serra, S., et al. (1985). Biopsychosocial characteristics of children who later murder: A prospective study. *American Journal of Psychiatry, 142*, 1161–1167.

Lewis, D. O., Pincus, J. H., Feldman, M., Jackson, L., & Bard, B. (1986). Psychiatric, neurological, and psychoeducational characteristics of 15 death row inmates in the United States. *American Journal of Psychiatry, 143*, 838–845.

Lewis, D. O., Pincus, J. H., Lovely, R., Spitzer, E., & Moy, E. (1987). Biopsychosocial characteristics of matched samples of delinquents and nondelinquents. *Journal of the American Academy of Child and Adolescent Psychiatry, 26*, 744–752.

Lewis, D. O., Yeager, C. A., Blake, P., Bard, B., & Strenziok, M. (2004). Ethics questions raised by the neuropsychiatric, neuropsychological, educational, developmental, and family characteristics of 18 juveniles awaiting execution in Texas. *Journal of the American Academy of Psychiatry and Law, 32*, 408–429.

Loftus, E. (2003). Our changeable memories: Legal and practical implications. *Nature Neuroscience, 4*, 232–233.

Lombroso, C. (1876). *L'Uomo delinquente in rapporto all'antropologia, alla giurisprudenza et alle discipline carcerarie* [The delinquent in anthropological, jurisprudential and medical views]. Milano: Hoepli.

Maguire, E. A., Gadian, D. G., Johnsrude, I. S., Good, C. D., Ashburner, J., Frackowiak, R. S. J., et al. (2000). Navigation-related structural change in the hippocampi of taxi drivers. *Proceedings of the National Academy of Sciences of the USA, 97*, 4398–4403.

Mark, V. H., & Ervin, F. R. (1970). *Violence and the brain.* New York: Harper & Row.

Markowitsch, H. J. (1992). *Intellectual functions and the brain. An historical perspective.* Toronto: Hogrefe & Huber Publs.

Markowitsch, H. J. (2003). Psychogenic amnesia. *NeuroImage, 20*, S132–S138.

Markowitsch, H. J., & Kalbe, E. (2007). Neuroimaging and crime. In S. Å. Christianson (Ed.), *Offender's memory of violent crime* (pp. 137–164). Chichester, UK: John Wiley & Sons.

Markowitsch, H. J., Calabrese, P., Würker, M., Durwen, H. F., Kessler, J., Babinsky, R., et al. (1994). The amygdala's contribution to memory – A PET-study on two patients with Urbach-Wiethe disease. *NeuroReport, 5*, 1349–1352.

Markowitsch, H. J., Kessler, J., Weber-Luxenburger, G., Van der Ven, C., & Heiss, W.-D. (2000). Neuroimaging and behavioral correlates of recovery from 'mnestic block syndrome' and other cognitive deteriorations. *Neuropsychiatry, Neuropsychology, and Behavioral Neurology, 13,* 60–66.

Meynert, T. (1867). Das Gesammtgewicht und die Theilgewichte des Gehirns in ihren Beziehungen zum Geschlechte, dem Lebensalter und dem Irrsinn, untersucht nach einer neuen Wägungsmethode an den Gehirnen der in der Wiener Irrenanstalt im Jahre 1866 Verstorbenen [The total and the partial weights of the brain and their relations to gender, age and insanity, investigated with a new weighing method on the brains of those who died in the Vienna lunatic asylum in the year 1866]. *Vierteljahresschrift für Psychiatrie, Psychologie und gerichtliche Medicin, 1,* 125–170.

Murray, J., & Farrington, D. P. (2005). Parental imprisonment: Effects on boys' antisocial behaviour and delinquency through the life course. *Journal of Child Psychology and Psychiatry, 46*, 1269–1275.

Popma, A., Vermeiren, R., Geluk, C. A., Rinne, T., van den Brink, W., Knol, D. L., et al. (2007). Cortisol moderates the relationship between testosterone and aggression in delinquent male adolescents. *Biological Psychiatry, 61*, 405–411.

Raine, A. (2001). Is prefrontal cortex thinning specific for antisocial personality disorder? *Archives of General Psychiatry, 58*, 402–403.

Raine, A., Lencz, T., Bihrle, S., LaCasse, L., & Colletti, P. (2000). Reduced prefrontal gray matter volume and reduced autonomic activity in antosocial personality disorder. *Archives of General Psychiatry, 57*, 119–127.

Raine, A., Lencz, T., Taylor, K., Hellige, J. B., Bihrle, S., Lacasse, L., et al. (2003). Corpus callosum abnormalities in psychopathic antisocial individuals. *Archives of General Psychiatry, 60*, 1134–1142.

Raine, A., Ishikawa, S. S., Arce, E., Lencz, T., Knuth, K. H., Bihrle, S., et al. (2004). Hippocampal structural asymmetry in unsuccessful psychopaths. *Biological Psychiatry, 55*, 185–191.

Saks, M. J., & Koehler, J. J. (2005). The coming paradigm shift in forensic identification science. *Science, 309*, 892–895.

Schiltz, K., Witzel, J., Northoff, G., Zierhut, K., Gubka, U., Fellmann, H., et al. (2007). Brain pathology in pedophilic offenders. *Archives of General Psychiatry, 64*, 737–746.

Sheline, Y. I., Gado, M. H., & Price, J. L. (1998). Amygdala core nuclei volumes are decreased in recurrent major depression. *NeuroReport, 9*, 2023–2028.

Sheline, Y. I., Gado, M., & Kraemer, H. (2003). Untreated depression and hippocampal volume loss. *American Journal of Psychiatry, 160*, 1516–1518.

Sie, M., & Wouters, A. (2008). The real challenge to free will and responsibility. *Trends in Cognitive Sciences, 12*, 3–4.

Siebert, M., Markowitsch, H. J., & Bartel, P. (2003). Amygdala, affect, and cognition: Evidence from ten patients with Urbach-Wiethe disease. *Brain, 126*, 2627–2637.

Smilansky, S. (2000). Free will: From nature to illusion. *Proceedings of the Aristotelian Society 101*, 71–95.

Spitz, R. A., & Wolf, K. (1946). Anaclitic depression. *Psychoanalytic Study of Children, 2*, 313–342.

Vermetten, E., Schmahl, C., Lindner, S., Loewenstein, R., & Bremner, J. D. (2006). Hippocampal and amygdalar volumes in dissociative identity disorder. *American Journal of Psychiatry, 163*, 630–636.

Vogt, O. (1929). 1. Bericht über die Arbeiten des Moskauer Staatsinstituts für Hirnforschung [First report on the work of the Moscow State Institute for Brain Research]. *Journal für Psychologie und Neurologie, 40,* 108–118.

Walter, H. (2002). Neurophilosophy of free will. In R. L. Kane (Ed.), *The Oxford handbook of free will* (pp. 565–577). Oxford: Oxford University Press.

Welt, L. (1888). Ueber Charakterveränderungen des Menschen infolge von Läsionen des Stirnhirns [On changes in character as a consequence of lesions of the frontal lobe]. *Deutsches Archiv für klinische Medicin, 42*, 339–390 (and 1 table).

Weiskrantz, L. (1973). Problems and progress in physiological psychology. *British Journal of Psychology, 64*, 511–520.

Winter, H., & Irle, E. (2004). Hippocampal volume in adult burn patients with and without posttraumatic stress disorder. *American Journal of Psychiatry, 161*, 2194–2200.

Witelson, S. F., Kigar, D. L., & Harvey, T. (1999). The exceptional brain of Albert Einstein. *Lancet, 353*, 2149–2153.

Yang, Y., Raine, A., Lencz, T., Bihrle, S., LaCasse, L., & Colletti, P. (2005). Volume reduction in prefrontal gray matter in unsuccessful criminal psychopaths. *Biological Psychiatry, 57*, 1103–1108.

Zeki, S., & Goodenough, O. (Eds) (2006). *Law and the brain.* Oxford: Oxford University Press.

NEUROCASE
2008, 14 (1), 7–14

Implications of fMRI and genetics for the law and the routine practice of forensic psychiatry

Harald Dressing, Alexander Sartorius, and Andreas Meyer-Lindenberg

Zentralinstitut für Seelische Gesundheit Mannheim, Universität Heidelberg, Mannheim, Germany

This review outlines recent neurobiological findings in humans relevant for the practice of law and forensic psychiatry. We focus on offenders with antisocial personality disorder and on sex offenders. In addition, the impact of risk polymorphisms in monoamine oxidase A (MAO-A), previously related to violence in interaction with the environment, on brain structure and function and on personality traits in healthy persons are presented. While increasing knowledge of functional and structural alterations provides a better understanding of the neurobiological underpinnings of delinquent behaviour, antisocial and violent behaviour arises from a complex pattern of biological, psychological, social and situational factors, precluding a stance of simple biological reductionism. Rather, optimal integration of neurobiological findings requires cooperation among many disciplines such as medicine, criminology, sociology, psychology, politics and neuroscience.

Keywords: Functional imaging; Antisocial personality disorder; Sexual offender; Aggression; Monoamine oxidase A.

INTRODUCTION

Brain imaging and genetic studies have increased our understanding of the neural basis underlying antisocial and sexual aggressive behaviour. Nevertheless, the neurobiological factors contributing to delinquent human behaviour remain poorly understood. This is perhaps unsurprising, as for any complex human trait, and particularly one that is partially defined by deviation from a societal norm that itself varies across time and space, an extensive causal network of biological, environmental, psychological and social factors must be considered, and no single factor in isolation is likely to play a major or even deterministic role. Consequently neurobiological studies in criminal offenders, although by themselves in the domain of clinical neuropsychiatric and psychological research, have given rise to a fairly controversial and hot tempered debate on the implications of these findings for the law and for forensic psychiatry. This debate touches on a discussion on the appropriateness of the concepts of free will, volitional control and criminal responsibility in the context of modern neurobiological results.

On the one hand, it has been argued that a neurobiological framework of deviant behaviour may eventually eliminate the concept of blame and responsibility (Sapolsky, 2004). Following this line of argument, the antisocial and violent behaviour of an offender may be conceptualized as a mechanistic consequence of brain dysfunction. In our view, this deterministic stance is biologically *a priori* unlikely (given the complexity of the mechanisms involved) and not supported by current empirical results. On the other hand it has been stated that human behaviour cannot adequately be described in terms of cause and effect. However, this view is also debatable (Dennett, 2003). Particularly for forensic psychiatric assessment, a psychological

Address correspondence to Professor Dr Andreas Meyer-Lindenberg, Zentralinstitut für Seelische Gesundheit, J 5, 68159 Mannheim, Germany. E-mail: a.meyer-lindenberg@zi-mannheim.de

http://www.psypress.com/neurocase   DOI: 10.1080/13554790801992800 approach that describes reasons, motives and intentions seems to be essential. From this point of view, human behaviour cannot be described adequately in physical terms. With regard to the wide spectrum of antisocial personalities, Kröber (2007) argues that these persons act as rational and competent citizens and their decision to behave against social norms should not be considered pathological. Finally, the medicolegal context of deviant behaviour further complicates the discussion as legal definitions and criteria do not identically map on either scientific or folk concepts of causation. Cautioned by these introductory remarks, in the following section some recent neurobiological research findings on antisocial personality disorder and paedophilia are outlined. In addition, results on neural mechanisms of genetic risk for impulsivity and personality traits in healthy volunteers are presented. The implications of these findings for the law and for the routine practice of forensic psychiatry will be discussed in the last section.

BRAIN IMAGING FINDINGS IN ANTISOCIAL INDIVIDUALS AND SEX OFFENDERS

Several brain imaging techniques have been made available in the past few decades. The following overview will focus particularly on functional magnetic resonance imaging (fMRI) findings in antisocial individuals and in paedophilic sex offenders, since informative recent results have been obtained in these conditions, the relevance of which for forensic psychiatric practice is well known.

The ability to control impulsivity and to exhibit socially adequate behaviour as well as the ability for moral and ethical judgement seems to be linked to the function of the prefrontal cortex (Bechara & van der Linden, 2005; Koenigs et al., 2007). However, it has to be kept in mind that the prefrontal cortex comprises fairly heterogeneous cerebral regions, among which the orbitofrontal cortex and the ventromedial cortex are of particular importance. Structural lesions of the prefrontal cortex may give rise to a syndrome of acquired psychopathy (Blair, 2003; Blair & Cipolotti, 2001; Meyers, Berman, Scheibel, & Hayman, 1992). Individuals with prefrontal brain lesions may develop symptoms, which according to Hare are typically found in psychopathy (e.g., loss of empathy, irresponsible behaviour, impulsivity). Psychopaths belong to a larger group of persons with antisocial personality

disorder and are characterized by the inability to have emotional involvement and by the repeated violation of the rights of others (Hare, 2003). These persons retain the cognitive capacity to distinguish right from wrong behaviour and, nonetheless, are incapable of regulating the appropriateness of their behaviour. In a quantitative MRI study, Raine, Lencz, Bihrle, LaCasse, and Colletti (2000) found at least 11% less prefrontal grey matter in men with antisocial personality disorder when compared with control subjects. The authors concluded that this is evidence of a structural brain deficit underlying the core symptoms of antisocial personality disorder. The fMRI technique offers even more sophisticated approaches to study the neurobiological corroboration of antisocial personality disorder and psychopathy. Since lesions in the frontolimbic circuitry may lead to behavioural manifestations designated as 'acquired sociopathy', functional underactivity in these circuits should predict behaviours qualitatively similar to those observed after manifest structural lesions in these regions. To analyse this hypothesis, 10 emotionally detached psychopaths were compared with 10 healthy control subjects in an fMRI study. Neuroimaging was performed during classic aversive differential delay conditioning. Based on the anatomical hypothesis the cingulate cortex, insular cortex, supplementary motor area, amygdala, OFC, and secondary somatosensory cortex were analysed. In this fMRI study, psychopaths displayed no significant activity in the limbic-prefrontal circuit during an aversive delay conditioning paradigm, while healthy controls showed enhanced differential activation in the limbic-prefrontal circuit during the acquisition of fear (Birbaumer et al., 2005). The authors concluded that the dissociation of emotional and cognitive processing may be a neural basis of the lack of anticipation of aversive events in psychopaths, and therefore these individuals will remain unable to learn from experience. Following this line of argument, punishment and incarceration might be a less efficacious way to deal with these offenders.

In another fMRI study, Schneider et al. (2000) investigated psychopaths by using an aversive classical conditioning paradigm during habituation, acquisition and extinction. In psychopaths, the amygdala showed a deactivation during unconditioned stimuli, which was interpreted as a correlate of their lack of response towards punishment. On the other hand, amygdala activation and activation of the dorsolateral prefrontal cortex was enhanced during acquisition, which was seen as an additional effort in those individuals trying to cope with the emotional deficit in the conditioning task. In a very recent study by Rilling et al. (2007), subjects with higher scores on psychopathy showed less amygdala activation during outcomes in which cooperation in a Prisoner's dilemma game was not reciprocated. They also displayed reduced activation within orbitofrontal cortex and dorsolateral prefrontal cortex when choosing to cooperate or to defect, respectively. In another fMRI study with six psychopaths, increased activation of right prefrontal regions and amygdala in response to images with negative contents was described (Müller et al., 2003).

Although the complex neurobiological underpinnings of psychopathy are incompletely understood at present, neurobiological research offers promising hypotheses. In particular, the data above suggest a dysfunction of the amygdala complex as part of a hypothetical neurobiological basis of the non-empathetic syndrome. Despite considerable public interest, research on neurobiological correlates of paedophilia is scarce. Our research group conducted the first fMRI study on paedophiles (Dressing, 2004; Dressing et al., 2001; Sartorius et al., in press; Tost et al., 2004). Our study included 10 male paedophilic subjects (exclusively attracted to boys), all convicted sex-offenders and sentenced to forensic psychiatric treatment along with 10 male heterosexual matched controls. We used a sexually non-explicit fMRI paradigm with images of men, women, boys or girls randomly embedded in neutral target/non-target geometrical symbols. While controls activated significantly less to pictures of children compared to adults, the activation profile was reversed in subjects with paedophilia, who exhibited significantly more activation to children than adults. The highest activation was observed for boys in the patient group, and for women in control participants. This study revealed an enhanced activation to children's pictures even in an incidental context and suggests the provocative hypothesis that a normally present mechanism for reduced emotional arousal for children relative to adults is reversed in paedophilia, suggesting a neural substrate associated with deviant sexual preference in this condition. Another study revealed decreased grey matter volume in the orbitofrontal cortex, the ventral striatum and the cerebellum indicating an association between frontostriatal morphometric abnormalities and paedophilia (Schiffer et al., 2007).

IMPACT OF FUNCTIONAL POLYMORPHISM IN MAOA ON BRAIN STRUCTURE, BRAIN FUNCTION AND PERSONALITY TRAITS IN HEALTHY INDIVIDUALS

If one looks at aggressive and sexual violent behaviour not from a categorical but from a dimensional perspective, studies on healthy individuals are of particular interest. In addition, scientific approaches that combine genetic, functional and structural analysis are promising to improve our understanding of socially deviant behaviour.

Since the most robust link between genetic variation and aggression has been shown for monoamine oxidase A (Caspi et al., 2002; Lesch & Merschdorf, 2000), a key enzyme for the catabolism of serotonin and other monoamines during neurodevelopment, the investigation of the impact of functional polymorphism in MAO-A on brain structure and function in healthy individuals can be seen as a prime example of such combined scientific approaches. Healthy individual with the low expression risk variant (MAO-A-L) showed, in an fMRI analysis during an emotional task, significantly increased activity in the left amygdala and decreased response in the subgenual and supragenual ventral cingulate cortex, left lateral orbitofrontal cortex and left insular cortex compared to individuals with the high expression variant of the MAO-A. There was also an effect on the brain structure. Individuals with the MAO-A-L variant showed a significant reduction in volume of the cingulate gyrus and bilateral amygdalae (Meyer-Lindenberg et al., 2006) (see Figure 1). Since amygdala was hyperreactive, while cingulated (and also orbitofrontal cortex) was hyporeactive in an emotionally face matching task, this suggested deficient prefrontal regulation in a circuit that has been linked to deficient fear extinction and impaired ability to deal with environmental stress (Pezawas et al., 2005). In addition, this study showed that men with the MAO-A-L variant revealed an increased activity in cerebral regions relevant for emotional memory and deficits in cerebral regions relevant for motor inhibition. In summary, these results reveal a neural mechanism that is associated with genetic risk for violent and impulsive behaviour. Since the sample comprised healthy volunteers, the data demonstrate that the genetic, functional and structural variations are compatible with normal health. Thus far the data should not be interpreted as showing a simple causal relationship between this genetic variant and aggression, as expected from a complex trait

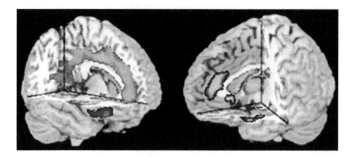

Figure 1. Allelic variation at the MAO-A u-VNTR affects brain structure and function. The left panel shows areas of decreased gray matter in MAO-A-L subjects (blue) and regions of increased gray matter in MAO-A-L men (red). The right panel shows aberrant functional activation during emotional face processing in MAO-A-L subjects (22). In both cases, amygdala and regions of the cingulated cortex are highlighted, suggesting a circuit for emotional regulation related to impulsivity and violence. This figure can be viewed in color on the Journal's website.

that is only partially determined by risk genes, none of which is likely to be of major effect, because interactions with other genes as well as social and psychological aspects have to be considered as significant factors contributing to violence in humans. Rather, the importance of these results lies in revealing neural circuits linked to violence by virtue of the prior genetic association found in large samples (Caspi et al., 2002), that is, the genetic association is used to extend knowledge about the neurobiology of a complex behaviour, a reverse genetics approach.

Thus far these data indicate a system of emotional dysregulation of limbic structures by the prefrontal cortex, they support concepts that impulsive violence has to be differentiated from the instrumental violence that is typical of a person with high psychopathy scores. Research on violence revealed at least two distinct types of violence. While impulsive violence refers to the satisfaction of destructive impulses and is linked to deficient inhibitory processes as well as to emotional arousal, instrumental violence is motivated by perceived personal gain (Tweed & Dutton, 1998). So far, scientific evidence supports the hypothesis that individuals who are prone to instrumental violence show a decreased activity of the amygdala and dysregulation of the prefrontal limbic neural network (Mobbs, Lau, Jones, & Frith, 2007). This implies that the increase in risk for violence linked to MAO-A is on account of the impulsive dimension.

Scientific studies that combine molecular genetic research and imaging techniques are also a promising approach in refining personality disorder diagnosis. Since personality disorder is an independent risk factor for violence and criminal recidivism (Coid, Hickey, Kahtan, & Zhang, 2007), the issue of

personality disorder diagnosis is of outstanding relevance in the context of violent and aggressive behaviour. A meta-analysis of more than 23,000 prisoners showed that 47% fulfilled criteria of antisocial personality disorder (Fazel & Danesh, 2002). Nevertheless, the categorical diagnosis of personality disorder yields major problems and may not fit to modern neurobiological findings. Therefore, a dimensional model that is based on behavioural continua and personality dimensions – e.g., harm avoidance, sensation seeking, reward dependence, etc. – may be more feasible (Livesley, 2005; Shedler & Westen, 2004; Yamaguta et al., 2006). Typical of the antisocial personality disorder are high scores on the dimension 'novelty seeking' and low scores on the dimension 'reward dependence' and 'harm avoidance' (Barnow et al., 2006). Since typical combinations of personality dimensions could be conceptualized as risk factors for aggressive behaviour it is important to understand their underlying neural mechanisms.

A combined genetic and imaging approach focused on MAO-A showed that male individuals with MAO-A-L had a dysregulated amygdala activation and increased functional coupling with ventromedial prefrontal cortex. The degree of this coupling correlated with high scores on the dimension 'harm avoidance' and low scores on the dimension 'reward dependence' (Buckholtz et al., 2007).

A very recent paper of Beaver et al. (2007) reports a gene × gene interaction effect of the DRD2 and the DRD4 dopamine receptor polymorphisms correlating with antisocial behaviour. While stating that such a correlation does not prove causation, it is still suggestive that a mechanism leading to dysfunctions in the mesolimbic (dopamine) reward systems may play a role in disrupting certain kinds of learning processes which are absent in psychopathy.

Although our understanding of the biological underpinnings of socially deviant behaviour is still poor, the presented data suggest at least some evidence for the complex interactions of neurobiological and sociobiographical factors that contribute to human violence. The consequences of such neuroscientific findings for the law and the routine practice of forensic psychiatry will be discussed in the following section.

IMPLICATIONS FOR THE LAW AND THE ROUTINE PRACTICE OF FORENSIC PSYCHIATRY

The idea that offenders suffering from a mental disorder must primarily be considered as ill and should therefore be exempted from punishment is of considerable antiquity and can be traced back to ancient Greek and Roman sources. It is grounded in the concept that mentally sane human beings have free will and are able to differentiate between right and wrong (Salize & Dressing, 2006). Only sane offenders should be punished. The concept of punishment seems to be a deeply rooted human trait and represents an important element in maintaining the stability of human societies (Fehr & Gächter, 2002). Punishment may be justified either by the idea of retribution or by the assumption of the future beneficial effects of this procedure (Greene & Cohen, 2004). The principle of retribution is based on the idea that punishment is aimed at giving the offender what he deserves based on his actions in the past. However, the prerequisite for retribution is that we are dealing with a mentally sane offender who possesses a free will and has chosen to violate the law. As indicated above, there is increasing evidence that impulsive aggression as well as sexual offending, and even the instrumental violence of psychopaths, is accompanied by significant neurobiological dysfunctions. However, it needs to be emphasized that even if differences in neurobiological parameters reach prespecified statistical levels of significance in group comparisons, this does not prove and usually not even suggest that these alterations play a major role in individual causation of deviant behaviour. A good example is again afforded by the cited studies of MAO-A in a large group of healthy controls, which yields significant findings

identifying a neural circuit linked to violent behaviour by virtue of genetic association, but does not imply a significant contribution in any of these subjects to violent behaviour. If this important caveat is ignored, or if future research should identify neural circuit dysfunction of demonstrable major impact on deviant behaviour, it could be argued that individuals who show similar neurobiological dysfunctions should be considered to be substantially impaired in their accountability and criminal responsibility. Following this line of argument, punishment of these persons that is based on the libertarian conception of free will and retributional thinking should not be considered to be an appropriate approach. However, even if so, punishment of criminal offenders – independent of their neurobiological status – could also be justified by the anticipated beneficial effects of this procedure for the offender himself as well as for society. With regard to the offender himself, punishment could be justified by the idea of prevention of further crimes, rehabilitation and reintegration. It has to be discussed to what extent the above-mentioned results of neurobiological research on aggression and sexual violence challenge these justifications for punishment. It could be argued that, at least for individuals diagnosed with psychopathy, this justification appears to be weak. It must be remembered that a key symptom of psychopathy is that afflicted persons do not learn from experience and do not anticipate the consequences of their behaviour. Birbaumer et al. (2005) presented a neurobiological mechanism for this deficit. As a consequence, the hypothesized beneficial effect of punishment in terms of primary deterrence with regard to the psychopath himself must be questioned. However, secondary deterrence with regard to the rest of society is also a worthy goal of punishment. The importance of secondary deterrence is that non-psychopathic members of society may learn not to violate the laws through the offender's example. So far, the strongest justification for punishment of individuals diagnosed as psychopathic is a utilitarian argument. It is the goal to protect society by the containment of these persons, since there is no question that violent and aggressive behaviour requires forceful interventions. It can be discussed controversially whether this containment should take place in prisons or in institutions under medical supervision. We must keep in mind that psychopathy currently must be considered as not treatable and that psychopaths are more likely than other people to re-offend. This may lead to the problematic requirement that psychopaths should not be punished, but that preventive and indefinite detention would be the correct way to handle them. If one looks at criminal responsibility, the argument could go in just the opposite way. One could argue that there is increasing evidence that psychopathy is a neurobiologically based mental disorder that reduces the capacity for self-control. Following this line of argument one could postulate that these people should not be punished and the sentence should be mitigated. It is evident that, with regard to the high risk of re-offending this would not be an acceptable solution for society. We have to keep in mind that similar problems have been discussed decades ago and some prior arguments are still relevant for the actual debate. In this context, Cleckley's publication 'The mask of sanity' is still outstanding. The following statement by him still seems to provide a very modern answer to some of the above-mentioned problems of how to deal with psychopaths: 'Despite traditional concepts and confusions, can we not conceive of a defect that seriously incapacitates and calls for restraining measures, without assuming that this defect necessarily absolves the subject from culpability and penalties of the law?' (Cleckley, 1988). Therefore, from several lines of reasoning, it seems questionable whether the above-cited neurobiological findings necessarily have the consequence that types of behaviour formerly considered as voluntary wrongdoing are now classed as disease. We stand at the beginning of an era that will probably see enormous growth in our knowledge of the neurobiological correlates of normal and socially deviant human behaviour. Functional brain abnormalities and genes that alone or in combination with environmental influences put persons at high risk to develop criminal behaviour will probably be identified. So far it is too early to give a final answer how society and the criminal justice system should respond to the ethical dilemmas that will arise from this knowledge (Appelbaum, 2005). The increasing knowledge of functional impairments provides a better understanding of the neurobiological underpinnings of delinquent behaviour. Optimal integration of neurobiological findings requires cooperation among many disciplines, such as medicine, criminology, sociology, psychology, politics and neuroscience. An intensive dialogue between neuroscientists, representatives of the legal profession, politicians, philosophers and other significant stakeholders is overdue.

Besides such more or less philosophical and sociological considerations on the implications of neurobiological research on legal regulations and the concept of criminal responsibility, some consequences for the routine practice of forensic psychiatry have to be discussed. One major issue in this context is the admissibility and probative value of fMRI images for the assessment of offenders. Of course these proposals must be considered in the light of the limitations which can be drawn from neuroimaging data of violent offenders. This is partly due to the fact that functional alterations in the amygdale–medial prefrontal cortex circuits have also been shown for example in patients with schizophrenia (Williams et al., 2007), posttraumatic stress disorder (Bryant et al., 2007) and borderline personality disorder (Minzenberg, Fan, New, Tang, & Siever, 2007). So far the use of expert testimony based on fMRI is scarce (Feigenson, 2006). Nevertheless, it can be expected that this technique will also be introduced into legal proceedings. However, the application of neurobiological research techniques in a context for which they were not intended raises some concerns (Garland, & Glimcher, 2006). One of the main concerns is that methodological limitations that are well known to researchers will not be realized by lawyers and judges. Laypersons may not realize that fMRI itself can not tell anything about the individual meaning of neural activation and the content of thoughts. In addition, the interpretation of fMRI results is to some degree individually variable and depends on the applied statistical methods. So far fMRI lacks, for the majority of applications of interest in the current context, diagnostic and predictive validity (Mobbs et al., 2007). However, some of these problems may be solved in the future. If a standardized set of genetic tests and fMRI paradigms for the forensic context were established, some problems of validity and reliability could be resolved. If brain imaging is applied to questions within clearly defined boundaries, and the applicability of the results to an individual context is clear and explicit, it has the potential to make useful contributions even for forensic psychiatric assessment in routine practice. Although fMRI and genetics are far from revolutionizing the legal system and the routine practice of forensic psychiatry, they will improve our understanding of important aspects of criminal behaviour and perhaps also change some strategies to deal with these problems in the courtroom and in prisons as well in forensic hospitals. Biological determinism is not the conclusion that can be drawn from the above-cited evidence, and personal accountability still remains an essential part of human existence (Henderson, 2005).

REFERENCES

Appelbaum, P. S. (2005). Psychiatry: Behavioral genetics and the punishment of crime. *Psychiatric Services, 56,* 25–27.

Barnow, S., Herpertz, S., Spitzer, C., Dudeck, M., Grabe, H. J., & Freyberger, H. J. (2006). Kategoriale versus dimensionale Klassifikation von Persönlichkeitsstörungen: Sind dimensionale Modelle die Zukunft? *Fortschritte der Neurologie und Psychiatrie, 74,* 706–713.

Beaver, K. M., Wright, J. P., Delisi, M., Walsh, A., Vaughn, M. G., Boisvert, D. et al. (2007). A gene x gene interaction between DRD2 and DRD4 is associated with conduct disorder and antisocial behavior in males. *Behavioural Brain Function, 22,* epub ahead of print.

Bechara, A., & van der Linden, M. (2005). Decision-making and impulse control after frontal lobe injuries. *Current Opinion in Neurology, 18,* 734–739.

Blair, R. J. (2003). Neurobiological basis of psychopathy. *British Journal of Psychiatry, 182,* 5–7.

Blair, R. J., & Cipolotti, L. (2001). Impaired social response reversal: A case of 'acquired sociopathy'. *Brain, 23,* 1122–1141.

Birbaumer, N., Veit, R., Lotze, M., Erb, M., Hermann, C., Grodd, W. et al. (2005). Deficient fear conditioning in psychopathy: A functional magnetic resonance imaging study. *Archives of General Psychiatry, 62,* 799–805.

Bryant, R. A., Kemp, A. H., Felmingham, K. L., Liddell, B., Olivieri, G., Peduto, A. et al. (2007). Enhanced amygdala and medial prefrontal activation during nonconscious processing of fear in posttraumatic stress disorder. An fMRI study. *Human Brain Mapping, 24,* epub ahead of print.

Buckholtz, J. W., Callicaott, J. H., Kolachana, B., Hariri, A. R., Goldberg, T. E., Genderson, M. et al. (2007). Genetic variation in MAOA modulates ventromedial prefrontal circuitry mediating individual differences in human personality. *Molecular Psychiatry,* epub ahead of print.

Caspi, A., McClay, J., Moffitt, T. E., Mill, J., Martin, J., Craig, I. W. et al. (2002). Role of genotype in the cycle of violence in maltreated children. *Science, 297,* 851–854.

Cleckley, H. C. (1988). *The mask of sanity* (5th ed.). Private printing for non-profit educational use. Copyright E. S. Cleckley, Augusta, GA. http://www.cassiopaea.org/

Coid, J., Hickey, N., Kahtan, N., & Zhang, T. (2007). Patients discharged from medium secure psychiatric services: Reconvictions and risk factors. *British Journal of Psychiatry, 190,* 223–229.

Dennett, D. C. (2003). The self as a responding-and-responsible-artifact. *Annals of the New York Academy of Sciences, 1001,* 39–50.

Dressing, H. (2004). Neurobiologische Befunde zur Pädophilie. *Nervenarzt, 75,* 264.

Dressing, H., Obergriesser, T., Tost, H., Kaumeier, S., Ruf, M., & Braus, D. F. (2001). Homosexual pedophilia and functional networks – An fMRI case report and literature review. *Fortschritte der Neurologie Psychiarie, 69,* 539–544.

Fazel, S., & Danesh, J. (2002). Serious mental disorder in 23000 prisoners: A systematic review of 62 surveys. *Lancet, 359,* 545–550.

Fehr, E., & Gächter, S. (2002). Altruistic punishment in humans. *Nature, 415,* 137–140.

Feigenson, N. (2006). Brain imaging and courtroom evidence: On the admissibility and persuasiveness of fMRI. *International Journal of Law in Context, 2*(3), 233–255.

Garland, B., & Glimcher, P. W. (2006). Cognitive neuroscience and the law. *Current Opinion in Neurobiology, 16,* 130–134.

Greene, J., & Cohen, J. (2004). For the law, neuroscience changes nothing and everything. *Philosophical Transactions of the Royal Society of London, 359,* 1775–1785.

Hare, R. D. (2003). *The Hare Psychopathy Checklist-Revised.* Toronto, Ontario: Multi-Health Systems.

Henderson, S. (2005). The neglect of volition. *British Journal of Psychiatry, 186,* 273–274.

Koenigs, M., Young, L., Adolphs, R., Tranel, D., Cushman, F., Hauser, M. et al. (2007). Damage to the prefrontal cortex increases utilitarian moral judgements. *Nature, 446,* 908–911.

Kroeber, H. L. (2007). The historical debate on brain and legal responsibility – revisited. *Behavioural Science Law, 25,* 251–261.

Lesch, K. P., & Merschdorf, U. (2000). Impulsivity, aggression, and serotonin: A molecular psychobiological perspective. *Behavioural Science Law, 18,* 581–604.

Livesley, W. J. (2005). Behavioural and molecular genetic contributions to a dimensional classification of personality disorder. *Journal of Personality Disorder, 19,* 131–155.

Meyer-Lindenberg, A., Buckholtz, J. W., Kolachana, B., Hariri, R., Pezawas, L., Blasi, G. et al. (2006). Neural mechanisms of genetic risk for impulsivity and violence in humans. *Proceedings of the National Academy of Sciences of the USA, 103,* 6269–6274.

Meyers, C. A., Berman, S. A., Scheibel, R. S., & Hayman, A. (1992). Case report: Acquired antisocial personality disorder associated with unilateral left orbital frontal lobe damage. *Journal of Psychiatry and Neuroscience, 17,* 121–125.

Minzenberg, M. J., Fan, J., New, A. S., Tang, C. Y., & Siever, L. J. (2007). Fronto-limbic dysfunction in response to facial emotion in borderline personality disorder: An event-related fMRI study. *Psychiatry Research, 155,* 231–243.

Mobbs, D., Lau, H. C., Jones, O. D., & Frith, C. D. (2007). Law, responsibility, and the brain. *PLoS Biology, 5,* e103.

Müller, J., Sommer, M., Wagner, V., Lange, K., Taschler, H., Roder, C. H. et al. (2003). Abnormalities in emotion processing within cortical and subcortical regions in criminal psychopaths: Evidence from a functional magnetic resonance imaging study using pictures with emotional content. *Biological Psychiatry, 15,* 152–162.

Pezawas, L., Meyer-Lindenberg, A., Drabant, E. M., Verchinski, B. A., Munzos, K. E., Kolchana, B. S. et al. (2005). 5 HTTLPR polymorphism impacts human cingulate-amygdala interactions: A susceptibility mechanism for depression. *Nature Neuroscience, 8*, 828–834.

Raine, A., Lencz, T., Bihrle, S., LaCasse, L., & Colletti, P. (2000). Reduced prefrontal grey matter volume and reduced autonomic activity in antisocial personality disorder. *Archives of General Psychiatry, 57*, 119–127.

Rilling, J. K., Glenn, A. L., Jairam, R., Pagnoni, G., Goldsmith, D. R., Elfenbein, H. A. et al. (2007). Neural correlates of social cooperation and noncooperation as a function of psychopathy. *Biological Psychiatry, 61*, 1260–1271.

Sapolsky, R. M. (2004). The frontal cortex and the criminal justice system. *Philosophical Transactions of the Royal Society of London, 359*, 1787–1796.

Salize, H. J., & Dressing, H. (2006). *Placement and treatment of mentally disordered offenders-Legislation and practice in the European Union.* Lengerich: Pabst Science Publishers.

Sartorius, A., Ruf, M., Kief, C., Demirakca, T., Bailer, J., Ende, G. et al. (in press). Abnormal amygdala activation profile in pedophilia. *European Archives of Psychiatry and Clinical Neuroscience.*

Schiffer, B., Peschel, T., Paul, T., Gizewski, E., Forsting, M., Leygraf, N. et al. (2007). Structural brain abnormalities in the frontostriatal system and cerebellum in pedophilia. *Journal of Psychiatric Research, 41*, 753–762.

Schneider, F., Habel, U., Kessler, C., Posse, S., Grodd, W., & Muller-Gartner, H. W. (2000). Functional imaging of conditioned aversive emotional responses in antisocial personality disorder. *Neuropsychobiology, 42*, 192–201.

Shedler, J., & Westen, D. (2004). Dimensions of personality pathology: An alternative to the five-factor model. *American Journal of Psychiatry, 161*, 1743–1754.

Tost, H., Vollmert, C., Brassen, S., Schmitt, A., Dressing, H., & Braus, D. F. (2004). Pedophilia: Neuropsychological evidence encouraging a brain network perspective. *Medical Hypotheses, 63*, 528–531.

Tweed, R. G., & Dutton, D. G. (1998). A comparison of impulsive and instrumental subgroups of batterers. *Violence and the Victim, 13*, 217–230.

Yamaguta, S., Suzuki, A., Ando, J., Ono, Y., Kijima, N., Yoshimra, K. et al. (2006). Is the genetic structure of human personality universal? A cross-cultural twin study from North America, Europe and Asia. *Journal of Personality and Social Psychology, 90*, 987–998.

Williams, L. M., Das, P., Liddell, B. J., Olivieri, G., Peduto, A. S., David, A. S. et al. (2007). Frontolimbic and autonomic disjunctions to negative emotion distinguish schizophrenia subtypes. *Psychiatry Research Neuroimaging, 155*, 29–44.

NEUROCASE
2008, 14 (1), 15–28

Neuropsychological and neural correlates of autobiographical deficits in a mother who killed her children

E. Kalbe,[1] **M. Brand,**[2] **A. Thiel,**[3] **J. Kessler,**[1] **and H. J. Markowitsch**[2]

[1]Department of Neurology, University Clinic Cologne, Cologne, Germany
[2]Department of Physiological Psychology, University of Bielefeld, Bielefeld, Germany
[3]Department of Neurology & Neurosurgery, SMBD Jewish General Hospital, McGill University Montreal, Montreal, Canada

We report a case of a delusional patient who had killed two of her children in an attempted 'extended suicide'. She was convinced of a genetic defect that caused autobiographical memory and emotional deficits and made life 'senseless'. Neuropsychological tests revealed dysfunctions in remembering emotional details of personal episodes and theory of mind. Water positron emission tomography (^{15}O) with a paradigm used in a former study by Fink et al. (1996) with healthy controls elicited abnormal activations during autobiographical memory retrieval characterised by a lack of prefrontal and limbic activity. We conclude that these imaging findings reflect neural correlates of the self-reported and objectified autobiographical dysfunctions. Furthermore, they indicate that beliefs or prejudices may have a major impact on the brain's processing of the personal past.

Keywords: Neuroscience; Crime; Autobiographical memory; Theory of mind; Positron emission tomography.

INTRODUCTION

Autobiographical-episodic memory is the highest memory system that consists of personal events with a clear relation to time, space and context; in Tulving's words, the conjunction of subjective time, autonoetic consciousness and the experiencing self represents episodic memories, which are typically emotionally toned (Tulving, 2002, 2005). Representing our personal past, autobiographical-episodic memory allows subjective time travel and is fundamental for building the feeling of one's own identity (Conway & Pleydell-Pearce, 2000). Accordingly, disorders of autobiographical memory usually lead to disastrous consequences of the individuals' lives as they lose their personal past and conse-

quently their personality (e.g., Markowitsch, 2003a, 2003b). Emotional deficits also commonly occur in those patients, as one function of autobiographical memory is to guide emotional behaviour on the bases of previous personal experiences having an emotional connotation (Kopelman, 2000; Serra, Fadda, Buccione, Caltagirone, & Carlesimo, 2007).

Retrograde amnesia for the personal past, covering either the whole life-span or distinct parts of the biography, can be caused by brain damage, predominantly if limbic and/or prefrontal structures are affected. In addition, psychic stress and traumas can also result in severe autobiographical memory disorders, a condition referred to as dissociative, psychogenic, or functional amnesia

This study was funded in part by the EC-FP6-project DiMI, LSHB-CT-2005-512146.

Address correspondence to E. Kalbe, Department of Neurology, University Hospital, Kerpener Str. 62, 50937 Cologne, Germany (E-mail: Elke.Kalbe@uk-koeln.de).

DOI: 10.1080/13554790801992735

(Brandt & van Gorp, 2006). In these conditions, autobiographical-episodic memory is most commonly affected. In addition, in some cases deficits in retrieving semantic information from their past or impairments in general semantic knowledge accompany the amnesia for personal episodes. A main characteristic of dissociative amnesia is that deficits occur in the absence of structural brain damage. Nevertheless, single-case reports point out that there may be functional brain changes, such as metabolic reductions in temporofrontal regions, as neural correlates of the memory disorders (Fujiwara et al., in press; Markowitsch, 1996a, 1996b, 1999a, 1999b; Markowitsch, Kessler, Van der Ven, Weber-Luxenburger, & Heiss, 1998; Reinhold, Kühnel, Brand, & Markowitsch, 2006; Sellal, Manning, Seegmuller, Scheiber, & Schoenfelder, 2002).

Patients with other psychiatric syndromes can also suffer from deficits in autobiographical memory or at least from specific changes in autobiographical remembering. For example, patients with depression retrieve less specific details of autobiographical memories (e.g., time and context information being less accurate) and show over-generalisation of remote autobiographical memory (review and meta-analysis in Van Vreeswijk & De Wilde, 2004). Instead of narrating temporally and contextually distinctive episodes, patients with depression tend to report categorical descriptions of summarised repeated occasions (Barnhofer, de Jong-Meyer, Kleinpass, & Nikesch, 2002; Williams et al., 2007). Patients with delusions (Kaney, Bowen-Jones, & Bentall, 1999) and schizophrenia also show less specific autobiographical memory. In the latter group, these alterations co-vary with deficits in emotional perspective-taking and theory of mind functions (Corcoran & Frith, 2003). The pattern of over-generalised memories in psychiatric diseases may be explained by dysfunctional integration of factual components of specific events from the personal past related to emotional experiences and self-related information. Potentially, abnormal functioning of the orbitofrontal cortex and therefore disruptions of the limbic-prefrontal connections may be neural correlates of the pattern described above, as shown, for instance, in patients with depression (Liotti, Mayberg, McGinnis, Brannan, & Jerabek, 2002). Furthermore, a dysregulation of the right brain has been postulated especially for cases with posttraumatic stress disorders (Schore, 1997).

In summary, autobiographical deficits, whether they are pronounced, as seen in patients with amnesia caused by brain damage or with dissociative amnesia, or whether they are more specific, as seen in depression or schizophrenia, can be linked to emotional dysfunctions and changes of the feeling of one's own self and may have distinct functional brain correlates.

It is also noteworthy that criminal offenders can have neuropsychological dysfunctions including deficits in emotional processing and memory (Deeley et al., 2006; Raine et al., 2005). In neuroscientific research, there is growing interest in investigating neural mechanisms underlying violent behaviour (Markowitsch & Siefer, 2007). Using modern brain imaging techniques, both structural and functional abnormalities were reported in individuals with antisocial personality disorder (Kiehl et al., 2001; Müller et al., 2003) as well as in murderers (Davidson, Putnam, & Larson, 2000; Raine, Stoddard, Bihrle, & Buchsbaum, 1998; Raine, Lencz, Bihrle, LaCasse, & Colletti, 2000). Typically, prefrontal and limbic brain regions are affected (Abbott, 2001; Brower & Price, 2001; Bufkin & Luttrell, 2005). However, it has not been reported yet that the personal burden due to autobiographical deficits or other cognitive dysfunction can lead to violent behaviour to oneself or others.

Here we report a so far unique case of a forensic delusional patient who had killed two of her own children in an act of 'extended suicide'. In her own view, she suffered from a genetic defect that led to an inability to build up an identity and feel emotionally connected to her past (which according to the theory described above reflects autobiographical memory dysfunction), and which consequently led to the same problems in her children. This train of thought was the self-reported reason for the offence. The aim of the current investigation was to reveal potential cognitive-mnestic and functional brain correlates of the patient's self-reported deficits. For this purpose, we administered an extensive neuropsychological test battery, and we measured the regional cerebral blood flow (rCBF) with positron emission tomography (PET) during an autobiographical memory paradigm that had been used in a former PET study with healthy control subjects (Fink et al., 1996).

CASE STUDY

Two years before our examination, the 49-year-old female, right-handed patient AA, a nurse, was admitted to a forensic clinic after having murdered two of her three children (9 and 10 years old) by first sedating them with doxepin at home and then drowning them in a river. She had then tried to commit suicide by cutting her arteries but was found early enough to be rescued. The offence was interpreted as an act of 'extended suicide'. According to the court expertise, which referred to an extensive psychiatric and neurological examination, she suffered from an affective disorder in the form of a long-lasting severe depressive episode with psychotic symptoms (F 32.3 according to ICD-10) and was thus found not guilty by reason of insanity. CT scans had shown no structural brain alterations.

AA's prominent psychotic symptom consisted of the delusion that she (and, consequently, also her three children) suffered from a genetic defect that made her life 'senseless'. She reported that this defect resulted in an inability to build up an identity and to feel emotionally connected to her past. She said that she was unable to feel emotions. Furthermore, she reported that she could not identify other people's emotions or intentions by interpreting their facial expressions or by grasping information from the intonation in speech. Although her behaviour in the clinic was regarded as 'social' and she took part in social events and even had the role of a 'spokesperson' in that setting, she also stated that she was unable to communicate with other people.

AA was convinced that death was the only way out and that it was right to kill her children. She also regarded the life of her oldest daughter, who was not at home when she committed the offence, as 'senseless' and believed that death would be better for her. Her descriptions of the offence were reported to have been unemotional.

At the time of our examination, 2 years after the offence, her mood was described as retained and not depressive any more. She had discontinued antidepressant medication about 1 year prior to our examination. Besides the delusion about her genetic defect, other psychotic symptoms were not detectable. There was no evidence of psychopathy according to the psychologist's evaluation and the Psychopathy Checklist-Revised (PCL-R, Hare, 2003) (score of 8 out of 40 points).

AA had great effort in finding objective evidence for her assumed defect, and she was convinced that so far no one had diagnosed her disease correctly. She attentively studied relevant popular scientific literature and had contacted a number of researchers, longing for an examination. It was also her initiative to contact one of the authors (HJM).

METHODS

Neuropsychology

An extensive neuropsychological test battery was administered to AA (Table 1). Besides an assessment of basic cognitive functions, instruments were used that test functions which, according to the patient's self reports, could be dysfunctional, that is autobiographical memory (concluded from the statement that she was unable to build an identity and to feel emotionally connected to her past), emotional processing, and theory of mind (inferred from reports of an inability to decode other people's emotions or intentions). It is noteworthy that the patient herself, as a layman, did not use the terms 'autobiographical memory' or 'theory of mind'.

In detail, the neuropsychological test battery included standardised instruments to assess the patient's general cognitive state (DemTect, Kalbe et al., 2004) and intelligence (Leistungsprüfsystem, Horn, 1983), executive functions (interference condition of the Word Colour Interference Test, Bäumler, 1985), verbal fluency using the letters F, A, and S (Strauss, Sherman, & Spreen, 2006), speed of information processing (conditions 'reading colour words' and 'naming colours' of the Word Colour Interference Test), and selective attention (d2 test, Brickenkamp & Zillmer, 1998).

Anterograde memory was assessed with the Wechsler Memory Scale-Revised (WMS-R, German version by Härting, Markowitsch, Neufeld, Calabrese, & Deisinger, 2000). Two further tests were used to determine the possible influence of the patient's emotional state on memory performance: The Affective Word Test (Fujiwara et al., in press) in which emotional and neutral words have to be recalled, and the emotional and neutral photographs (e.g., those used by von Cramon, Markowitsch, and Schuri, 1993). In this latter task, 40 out of 80 photographs of various scenes have to be recognised after a 15-min delay. Half of the photographs had been rated as particularly touching emotionally by 20 normal subjects.

Autobiographical memory was assessed with the Bielefeld Autobiographical Memory Inventory (Fast & Markowitsch, in press), a semistructured interview which is based on the technique proposed by Kopelman, Wilson, and Baddeley (1989) and in which autobiographical facts and autobiographical episodic memory from specified time intervals of the whole life span have to be recalled and evaluated regarding vividness, detailedness, and emotionality.

TABLE 1
Results of the neuropsychological test battery

Domain	Test/Instrument	AA's performance	Reference score
Cognitive state/ Intelligence	DemTect	16	15.3 (2.8)
	Leistungsprüfsystem (LPS-K)	PR: 54	PR: 50
Executive functions, attention and speed of information processing	Verbal Fluency: FAS-test, sum of words	PR: 80	PR: 50
	Word Colour Interference Test: reading colour words	T: 56	T: 50 (20)
	colour naming	T: 55	T: 50 (20)
	interference trial	T: 48	T: 50 (20)
	Selective Attention Test (d2): total minus errors	PR: 46	PR: 50
Anterograde memory	WMS-R (indices): verbal memory	95	100 (15)
	visual memory	94	100 (15)
	general memory	94	100 (15)
	attention/concentration	106	100 (15)
	delayed recall	116	100 (15)
	Affective Word Test: affective judgements (errors)	1	1.3 (1.3)
	free recall	6	4.1 (1.8)
	recognition	13	13.0 (1.6)
	false positives	0	1.4 (1.4)
	Recognition of Emotional and Neutral Photographs:		
	emotional	**64%**	95%
	neutral	**50%**	88%
Retrograde memory	Bielefeld Autobiographic Memory Inventory:		
	semantic	80%	90 (11)%
	episodic: free	**53%**	89 (11) %
	episodic: details	**20%**	85 (16) %
	period of life: pre-school age	**50%**	90 (12) %
	primary school age	**55%**	94 (15) %
	secondary school age	**55%**	934 (6) %
	early adulthood (up to 35 years)	**36%**	64 (5) %
	latest past	**23%**	91 (15) %
	total	**43%**	96 (9) %
	Kieler Remote Memory Test	43%	cut-off: >40%
	Famous Faces Test:		
	verbal, recognition	90%	94 (8)
	verbal, semantic information	90%	89 (14)
Affective processing	Ekman & Friesen: naming	11	11 (1)
	multiple choice	**16**	22 (2)
	Tübinger Affect Battery:		
	facial identity discrimination	100%	99 (3)
	facial affect discrimination	93%	92 (6)
	facial affect naming	93%	95 (6)
	facial affect selection	100%	97 (4)
	facial affect matching	**73%**	94 (5)
	nonemotional prosody discrimination	**73%**	100 (0)
	emotional prosody discrimination	100%	100 (0)
	name the emotional prosody	93%	95 (8)
	match emotional prosody to an emotional face	93%	96 (7)
Theory of mind	Reading-the-Mind-in-the-Eyes-Test (max. score = 18)	**12**	15 (2)
	Happé stories:		
	Theory of Mind Stories (max. score = 16)	**10**	15 (1)
	Control 'Physical' Stories (max. score = 16)	10	12 (2)
	Multiple-Choice-Theory-of-Mind-Test		
	(max. score = 16)	**9**	14 (2)

Impaired scores in AA's performance (deviation of more than 1 *SD* from reference data) are indicated in bold.

Further tests for remote memory were the Kieler Remote Memory Test (Kieler Altgedächtnistest, Leplow, Blunck, Schulze, & Ferstl, 1993; Leplow, Dierks, Merten, & Hänsgen, 1997) which assesses knowledge of famous events, and the Famous Faces Test (Fast, Fujiwara, & Markowitsch, in preparation) in which famous persons from politics, sports, and cultural life have to be recognised.

Affective processing was tested with a set of 28 black and white photographs taken from Ekman and Friesen (1975) which show facial emotional and neutral expressions and which had to be described first and then matched to corresponding terms in a multiple choice design (anger, disgust, fear, joy, sadness, surprise, or neutral). Furthermore, the Tübinger Affect Battery (Breitenstein, Daum, Ackermann, Lütgehetmann, & Müller, 1996), a German version of the Florida Affect Battery (Bowers, Blonder, & Heilman, 1998), was used. It includes two sets of tasks testing the ability to process facial affect and emotional prosody, respectively. Finally, theory of mind, i.e., the ability to infer other people's mental states such as emotions, thoughts and intentions, was administered with the Reading-the-Mind-in-the-Eyes-Test (Baron-Cohen, Wheelwright, Hill, Raste, & Plumb, 2001, German version by Kalbe et al., unpublished material), with a story comprehension task including theory of mind stories and control (non-theory of mind) stories (German translations of the material introduced by Happé, 1994) as well as with the Multiple-Choice-Theory-of-Mind-Test (Kalbe et al., unpublished material) in which character's thoughts in short scenes have to be inferred in a multiple choice paradigm.

Positron-Emission-Tomography (PET)

To find possible neural correlates of the episodic autobiographical memory dysfunctions that were derived from AA's reports and were confirmed by neuropsychological tests (see below), an autobiographical memory paradigm was chosen that had been used by a former study by Fink et al. (1996) with healthy controls. AA's individual results could thus be compared to those of historical controls scanned according to exactly the same protocol.

The PET design included 12 sequential measurements of relative regional cerebral blood flow (rCBF). It consisted of three study conditions: one control condition (REST [A]) with eyes closed and no auditory or visual stimulation, and two experimental conditions: in the FICTITIOUS (B) condition, auditory sentences were presented that contained episodic autobiographical information of a person our patient AA did not know, but which had been introduced to her about 1 h before PET scanning. In the AUTOBIOGRAPHICAL (C) condition, sentences were read to AA that contained episodic information taken from her own life but not referring to the offence. All episodes were brief and significant events of the other person's life (FICTITIOUS, e.g., 'With my French exchange student I mixed cocktails' or 'My first car was a VW beetle') and her own life, respectively (AUTOBIOGRAPHICAL, e.g., 'From my Russian pen pal I received a small parcel' or 'A cat that frequently crept around our garden became our pet'). AA received the instruction to imagine what happened to the person in the described situations (FICTITIOUS) or to imagine what happened to herself in the described situations (AUTOBIOGRAPHICAL). AA's episodes were partly taken from the patient's history and partly referred to information she had given in the Autobiographical Memory Inventory. The fictive episodes of the other person were matched to those taken from AA's life concerning topics and emotionality. The presentation order of the study condition was ABCCBAACBBCA.

Relative rCBF was measured by recording the regional distribution of cerebral radioactivity after the intravenous injection of ^{15}O-labeled water. Measurements were performed on an ECAT-EXACT HR scanner (Siemens Medical Systems, Erlangen, Germany) in 3D-Mode. After correction for measured attenuation and scatter, 12 images consisting of 47 transaxial slices with a thickness of 3.125 mm were reconstructed using a filtered back-projection algorithm with a Hanning-filter (cut-off frequency of 0.4 pixels/cycle). For data analysis, images were ratio-normalised for differences in global CBF, filtered with a spherical Gaussian kernel of 8 mm FWHM, and difference images of the contrast AUTOBIOGRAPHICAL–REST, FICTITIOUS–REST and AUTOBIOGRAPHIC–FICTITIOUS were computed. Significance was assessed on z-transformed images using a global variance estimate from balanced noise images as previously described (Thiel et al., 2001). Voxels with z-scores greater than 3 or less than –3 were regarded as significantly activated or deactivated. Z-score images were transformed to the MNI152-standard brain using FSL (FMRIB, Oxford) and activation/deactivations are reported in Talairach coordinates.

RESULTS

Neuropsychology

An overview of the neuropsychological test results is given in Table 1. AA was well adapted and cooperative throughout the examination. She was well oriented to time and place. Her cognitive performance was within or above average in tests for the general cognitive state, intelligence, executive functions, attention, and speed of information processing.

Anterograde memory performance as assessed with the WMS-R and also the Affective Word Test was in the normal range. Remote memory as tested with the Kieler Remote Memory Test and the Famous Faces Test was unaffected in AA. In contrast, marked deficits were evident in autobiographical memory. The Autobiographical Memory Inventory revealed that, while semantic autobiographical memory was only slightly below the average, the recall of personal episodes was thoroughly disturbed with a temporal gradient towards the present. The classification of episodes remembered revealed that AA mostly remembered general, recurring events, i.e., events that recurred over a longer period of time such as 'With our athletic sports group we regularly participated in contests' or 'My friend lived with her grandparents and we used to play at their big kitchen table'. These general events and also the few single events AA remembered could not be described in detail. She also reported that she could not visualise these memories at all and that remembering them did not arouse any emotional involvement.

Contradicting results were observed regarding affective processing in AA. While the majority of subtests of the Tübinger Affect Battery and also naming of facial expressions using the pictures from Ekman and Friesen (1975) were within the normal range, AA scored below average both in the facial affect matching task of the Tübinger Affect Battery and in the comparable condition using the Ekman pictures (matching facial expressions to corresponding terms of emotion). Significant disturbance of the ability to infer other people's mental states were observed in all used theory of mind tasks. It is noteworthy that AA's performance in the control 'physical' stories of the Happé story task was unaffected.

PET

Activation and deactivation patterns in patient AA

Contrasting the two activation conditions, AUTO-BIOGRAPHICAL–FICTITIOUS, revealed signi-ficant activations within the right posterior and left anterior cingulate gyrus as well as anterior part of the right insular cortex (Table 2 and Figure 1). Furthermore, several regions were deactivated significantly, i.e., the right anterior cingulate gyrus, the right medial and the lateral orbitofrontal cortex, the left inferior frontal gyrus and the left angular gyrus.

When the REST condition was subtracted from the AUTOBIOGRAPHICAL condition, activations were found in the left and right superior temporal gyrus as well as the temporal and frontal poles. Deactivations were demonstrated in the right anterior and middle cingulate gyrus and primary and visual association cortices of both hemispheres.

In the FICTITIOUS minus REST contrast, inferior frontal and lateral temporal regions bilaterally as well as the left parietal association cortex were activated. The right sided primary visual cortex and the middle part of the cingulate gyrus of both hemispheres were deactivated. Furthermore, the left anterior and right posterior cingulate gyrus were deactivated. These latter deactivations most likely lead to the activations of these areas found in the AUTOBIOGRAPHICAL minus FICTITIOUS contrast.

Comparison to normal healthy controls

Compared to the historical healthy controls (Fink et al., 1969) there were no significant activations found in right medial temporal lobe structures (e.g., hippocampal formation, amygdala) for the contrast AUTOBIOGRAPHICAL–FICTITIOUS, whereas the left-sided activations observed in our patient were not reported in healthy control subjects.

In the contrast AUTOBIOGRAPHICAL–REST, again wide-spread activations of the left hemisphere in patient AA were found which were not observed in normal controls. Concerning the deactivations, the same occipital pattern of task related CBF-decrease was observed as in controls but the deactivation of the right cingulate was only seen in patient AA.

Differences were also seen for the contrast FICTITIOUS–REST. In healthy controls the parietal regions, which were activated in patient AA, were found to be inactivated. In contrast, the posterior cingulate region was activated in controls and deactivated in AA. The deactivation of the occipital cortex was observed in our patient as in controls.

TABLE 2
Talairach coordinates of active and inactive regions in different contrasts

Condition	Region	Side	x	y	z	Z-score
Autobiographical–Ficticious						
Active	Anterior cingulate gyrus	L	−11	44	14	3.37
	Posterior cingulate gyrus	R	5	−46	21	3.54
	Anterior insula	R	25	25	5	3.43
Inactive	Anterior cingulate gyrus	R	6	33	31	−3.16
	Medial orbitofrontal cortex	R	15	55	−9	−4.41
	Lateral orbitofrontal cortex	R	46	36	−4	−4.00
	Medial frontal gyrus	L	−35	31	12	−3.78
	Angular gyrus	L	−48	−43	49	−3.96
Autobiographical–Rest						
Active	Frontal pole	R/L	−1	58	30	4.27
	Anterior insula/temporal pole	R	29	29	4	4.99
	Frontal inferior gyrus/ temporal pole	L	−50	25	1	4.76
	Superior temporal gyrus	R	53	−21	6	3.85
	Superior temporal gyrus	L	−58	−19	3	4.46
Inactive	Anterior cingulate gyrus	R	6	37	11	−4.41
	Middle cingulate gyrus	R	2	−10	39	−3.73
	Primary visual cortex	L/R	−8	−89	14	−3.16
	Visual associative cortex	R	28	−62	−1	−3.27
	Visual associative cortex	L	−34	−85	−5	−3.95
Ficticious–Rest						
Active	Frontal superior gyrus	R	7	33	50	3.53
	Inferior frontal gyrus	R	44	34	−1	3.79
	Inferior frontal gyrus	L	−42	22	3	4.37
	Superior temporal gyrus	R	56	−32	5	3.18
	Superior temporal gyrus	L	−62	−15	1	4.13
	Parietal associative cortex	L	−49	−43	48	4.89
Inactive	Anterior cingulate gyrus	L	−7	32	12	−3.00
	Posterior cingulate gyrus	R	4	−45	25	−3.30
	Middle cingulate gyrus	R/L	0	−3	35	−3.36
	Primary visual cortex	R	5	−70	15	−3.00

All activations and deactivations significant at $p < .05$, corrected for multiple comparisons across the whole brain volume. For each region, the coordinates in standard stereotactic space are given, referring to the maximally activated focus within an area of activation or deactivation, respectively, as indicated by the highest Z-score. x is the distance (mm) to right (R; +) or left (L; −) of the midsagittal plane; y is the distance anterior (+) or posterior (−) to vertical plane through the anterior commissure; z is the distance above (+) or below (−) the inter-commissural (AC–PC) plane.

DISCUSSION

We here describe the case of a forensic delusional patient AA who killed two of her own children in an act of extended suicide. The patient herself reported that her main dysfunction was not building an identity or feeling emotionally connected to her past (in the author's view describing episodic autobiographical memory dysfunction, see Conway & Pleydell-Pearce, 2000) as well as an inability to feel and decode emotions (interpreted as theory of mind dysfunction) and to communicate. Our investigation revealed two main findings: (1) the neuropsychological profile is in accordance with the self-reported deficits. She suffered from selective autobiographical episodic memory changes. In detail, there was a lack of specificity and emotionality in remembering episodes from her past. Moreover, theory of mind deficits were observed. (2) Specific brain activation patterns linked to autobiographical memory retrieval as examined with PET indicate potential neural correlates of her self-reported dysfunctions.

Referring to the neuropsychological pattern, AA had intact retrograde memory both for general semantic information as well as facts of her biography. She also retrieved episodes from her life. However, the events retrieved usually had more general character in the way that they were not single episodes but happened repeatedly (e.g., 'Every morning on the way to school..'). Thus, they do not reflect what Tulving (2002, 2005) is referring to

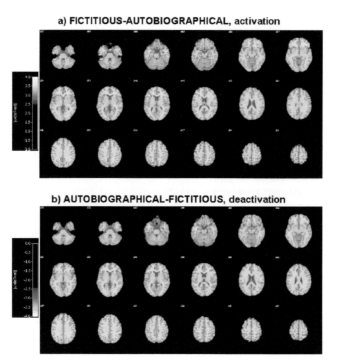

Figure 1. $[^{15}O]H_2$ PET images of patient AA with activations (a) and deactivations (b) in the main contrast AUTOBIOGRAPHICAL–FICTITIOUS.

as 'episodic memories' which have a clear relation to a specific time and a singular locus. Most importantly, she was unable to report affective connotations of autobiographical episodes and did not have a feeling of her own self when retrieving events. This was even evident when she reported the act of killing her children. The report included the temporal order of acting on the day of the offence, but no emotional approach was perceived. These circumstances also indicate that she did not vividly remember her biography as defined by Tulving, who proposed that episodic memories are emotionally toned and self-related. Accordingly, we suggest that AA suffered from a specific form of autobiographical memory dysfunction.

As outlined in the introduction, deficits in autobiographical remembering including the emotional connotation of episodes are the main symptom of dissociative amnesia (e.g., Brandt & van Gorp, 2006; Kopelman, 2000; Kritchevsky, Chang, & Squire, 2004; Markowitsch, 2003b). Therefore, one might hypothesise that AA's deficits reflect symptoms of this kind of dissociative condition. On the other hand, the mnestic profile revealed is not directly comparable to the pattern typically found in patients with this form of amnesia. These patients commonly 'forget' a whole life span or episodes from the whole life including the sense of

their own identity (Fujiwara et al., in press). In fact, our patient AA reported that she had no sense for her own self. Nevertheless, she could report episodes from all time periods of her life. Therefore, we do not think that she matches with the condition of dissociative amnesia, at least not in its typical form.

Recalling more general than specific events of the personal past has been referred to as 'over-generalised memory'. This phenomenon is often accompanied by less frequent retrieval of episodic-specific emotional connotations, as seen for example in patients with depression (Barnhofer et al., 2002; de Decker, Hermans, Raes, & Eelen, 2003; Van Vreeswijk & De Wilde, 2004; Williams, 1996), persecutory delusions (Kaney et al., 1999), or schizophrenia (Riutort, Cuervo, Danion, Peretti, & Salame, 2003; Sonntag et al., 2003). These groups of patients also frequently show theory of mind dysfunctions (Harrington, Siegert, & McClure, 2005b; Lee, Harkness, Sabbagh, & Jacobson, 2005). It is noteworthy that it has been proposed that there is a relationship between a tendency to over-generalised memory and theory of mind reductions (Corcoran & Frith, 2003). This might explain the convergent neuropsychological pattern that was observed in our patient AA, who suffered from theory of mind abnormalities beyond the specific autobiographical memory alterations mentioned above. Potentially, deficits in inferring others' mental states, especially emotions, can compromise retrieval of specific events, as it is difficult to develop a feeling of one's own emotional state without being able to access main features of social interactions. In other words, attributions to others can serve as cues for remembering contextual details of an event (for further discussion see Corcoran & Frith, 2003). Additionally, the theory of mind reductions may also contribute to AA's delusional symptoms, given that this specific link has been shown in patients with paranoid delusions (Craig, Hatton, Craig, & Bentall, 2004; Harrington, Langdon, Siegert, & McClure, 2005a).

As potential neural correlates of AA's autobiographical memory complaints (circumscribed as the inability to build an identity and feel connected to her past), the PET investigation revealed a lack of activation in brain structures that have previously been linked to autobiographical memory retrieval. It is noteworthy that the lack of a direct control group is a limitation of our study. However, AA's results could be compared to historical control subjects of a former study (Fink et al., 1996) which used exactly the same PET paradigm.

The main deviation of AA's activation patterns compared to our historical control subjects and further studies were consistent in less recruitment of right hemisphere regions and more activation in left-sided areas, which are not observed in normal controls. For instance, there are several studies indicating that the prefrontal cortex is crucially engaged in retrieval processes. However, in patient AA, no prefrontal activations were found while she attempted to retrieve episodes from her past. By contrast, the medial and lateral orbitofrontal cortices were even deactivated in the AUTOBIO-GRAPHICAL minus FICTITIOUS contrast. This pattern potentially reflects AA's specific autobiographical memory dysfunction, as there is convergent evidence for the involvement of both the lateral and the medial orbitofrontal cortex section in the retrieval of the emotional tone of personal episodes (Brand & Markowitsch, 2006; Cabeza et al., 2004; Conway et al., 1999; Keedwell, Andrew, Williams, Brammer, & Phillips, 2005; Levine, 2004; Maguire, Henson, Mummery, & Frith, 2001; Markowitsch, Vandekerckhove, Lanfermann, & Russ, 2003; Ryan et al., 2001).

Concerning the lateral orbitofrontal cortex, there is also evidence for its engagement in discriminating lies from truth (Langleben et al., 2005;

Markowitsch et al., 2000; Spence et al., 2001) and in telling lies and deceptions. Our patient's deactivation in this region further emphasizes possible difficulties in processing details of episodic memories and in using these details for discriminating real and fictitious episodic information.

Beyond the prefrontal regions, structures of the medial temporal lobe, e.g., the hippocampal formation and the amygdala, were also not activated during AA's autobiographical memory retrieval attempts. Previous studies with healthy individuals consistently found these structures activated when remembering personal episodes (Addis, Wong, & Schacter, 2007; Cabeza & St. Jacques, 2007; Fink, 2003; Fink et al., 1996; Maguire & Frith, 2003; Piefke, Weiss, Zilles, Markowitsch, & Fink, 2003; Piefke, Weiss, Markowitsch, & Fink, 2005; Piefke & Fink, 2005; Vandekerckhove, Markowitsch, Mertens, & Woermann, 2005; Viard et al., 2007).

Some activations seen in AA conformed with findings in healthy control subjects. In the main contrast (AUTOBIOGRAPHICAL–FICTITIOUS), activations were found within the cingulate gyrus (posterior on the right and anterior on the left side) as well as in the anterior part of the right insula. In previous studies, the anterior cingulate gyrus was found to be involved in a number of emotional and motivational processes, but also in attention and executive functions such as set-shifting and cognitive flexibility (Allman, Hakeem, Erwin, Nimchinsky, & Hof, 2001; Kondo, Osaka, & Osaka, 2004; Lie, Specht, Marshall, & Fink, 2006; Otsuka, Osaka, Morishita, Kondo, & Osaka, 2006; Shima & Tanji, 1998). It also plays a crucial role in social cognition (Frith, 2002, 2007). Furthermore, it is fundamentally involved in a network underlying episodic memory encoding and retrieval (e.g., Brand & Markowitsch, 2003; Markowitsch, 1995, 2000; Piefke et al., 2003; Svoboda, McKinnon, & Levine, 2006; Vandekerckhove et al., 2005). Beyond this, the anterior section of the cingulate gyrus is also associated with the detection of conflicts and resulting erroneous responses (Bush, Luu, & Posner, 2000; Carter et al., 1998; Gehring & Knight, 2000; Hester, Fassbender, & Garavan, 2004; Luu, Flaisch, & Tucker, 2000; van Veen & Carter, 2006; Yeung, Holroyd, & Cohen, 2005). Given that the activation design used in our patient study comprised items which were either autobiographical or fictitious, the task required the patient to detect which category each item belonged to. Therefore, the activation of the anterior cingulate gyrus potentially reflects processing this kind of cognitive conflict, as demonstrated by previous studies mentioned above.

Regarding the activation in the right posterior cingulate gyrus and precuneus, this finding might indicate that our patient did recognise the sentences describing personal episodes as belonging to her own past. The precuneus has been found to be part of a network for self-referential processing or self reflective consciousness (e.g., Cavanna & Trimble, 2006; Johnson et al., 2002; Kjaer, Nowak, & Lou, 2002; Uddin, Kaplan, Molnar-Szakacs, Zaidel, & Iacoboni, 2005; Vogeley et al., 2004). Activation of this region was also seen in studies of episodic and autobiographical memory retrieval (e.g., Lou et al., 2004; Maddock, Garrett, & Buonocore, 2001; Piefke et al., 2003; Sugiura, Shah, Zilles, & Fink, 2005; Svoboda et al., 2006). Furthermore, it may be involved particularly in the retrieval of visualisable information (Fletcher et al., 1995; Fletcher, Shallice, Frith, Frackowiak, & Dolan, 1996), though this is controversial.

Nevertheless, one has to notice that both the anterior and the posterior part of the cingulate gyrus were activated in the contrast AUTOBIO- GRAPHICAL minus FICTITIOUS but deactivated in the contrast FICTITIOUS minus REST. This pattern suggests that the cognitive conflict of detecting true and fictitious information, as argued above, did not occur – or at least not as strongly – while processing fictitious episodes. This means that the patient most likely had no problem to decide whether an event was fictitious or not and that performing the FICTITIOUS condition did not require self-referential processing. On the other hand, recognising a true event as belonging to her own biography was more directly tapping into processing self relevant information and may thus have been more effortful compared to processing the fictitious events. However, this point remains speculative, as no behavioural measure was included in the paradigm. The result that the cingulate gyrus was not activated in the AUTOBIOGRAPHICAL minus REST contrast may result from self-referential processing also in the REST condition; a phenomenon which is also frequently referred to as 'default mode' (Greicius, Krasnow, Reiss, & Menon, 2003; Hampson, Driesen, Skudlarski, Gore, & Constable, 2006).

One interesting question is whether the psychopathological symptoms in our patient, namely delusions, were related to the neural alterations described above. Functional imaging studies examining autobiographical memory in delusional patients are sparse. However, existing data have not shown changes in activation patterns which are similar to what we observed in our patient. For example, Blackwood et al. (2004) found abnormalities of cingulate gyrus activation (which was not abnormal in AA) while determining self-relevance in schizophrenic patients with active persecutory delusions. It may thus be assumed that the delusions themselves were not related to the changes we observed in the present study. However, since our paradigm or similar ones so far have not been used in patients with delusions, the exact relationship between deficits in autobiographical memory and delusions remains to be further characterised. This endeavour, however, is beyond the scope of this case report.

In summary, our patient suffered from specific neuropsychological deteriorations covering the ability to remember her own past emotionally and vividly, as well as deficits in inferring other people's mental states. In addition, functional imaging elicited an abnormal activation pattern characterised by a lack of prefrontal and limbic, especially medial temporal engagement in episodic memory retrieval. To conclude, this study demonstrates that self-reported autobiographical dysfunctions accompanied by theory of mind deficits – which according to our patient were the reason for the offence in order to prevent her children from a 'senseless life' – can have neuropsychological and functional brain correlates.

REFERENCES

Abbott, A. (2001). Into the mind of a killer. *Nature, 410,* 296–298.

Addis, D. R., Wong, A. T., & Schacter, D. L. (2007). Remembering the past and imagining the future: Common and distinct neural substrates during event construction and elaboration. *Neuropsychologia, 45,* 1363–1377.

Allman, J. M., Hakeem, A., Erwin, J. M., Nimchinsky, E., & Hof, P. (2001). The anterior cingulate cortex: The evolution of an interface between emotion and cognition. *Annals of the New York Academy of Sciences, 935,* 107–117.

Barnhofer, T., de Jong-Meyer, R., Kleinpass, A., & Nikesch, S. (2002). Specificity of autobiographical memories in depression: An analysis of retrieval processes in a think-aloud task. *British Journal of Clinical Psychology, 41,* 411–416.

Baron-Cohen, S., Wheelwright, S., Hill, J., Raste, Y., & Plumb, I. (2001). The 'Reading the Mind in the Eyes' Test revised version: A study with normal adults, and adults with Asperger syndrome or high-functioning autism. *Journal of Child Psychology and Psychiatry, 42*, 241–251.

Bäumler, G. (1985). Farbe-Wort-Interferenz-Test (FWIT). Göttingen: Hogrefe.

Blackwood, N. J., Bentall, R. P., FFytche, D. H., Simmons, A., Murray, R. M., & Howard, R. J. (2004). Persecutory delusions and the determination of self-relevance: An fMRI investigation. *Psychological Medicine, 34*, 591–596.

Bowers, D., Blonder, L., & Heilman, K. (1998). *The Florida Affect Battery*. Gainesville, FL: University of Florida Brain Institute.

Brand, M., & Markowitsch, H. J. (2003). The principle of bottleneck structures. In R.H. Kluwe, G. Lüer, & F. Rösler (Eds), *Principles of learning and memory* (pp. 171–184). Basel: Birkhäuser.

Brand, M., & Markowitsch, H. J. (2006). Memory processes and the orbitofrontal cortex. In D. Zald, & S. Rauch (Eds), *The orbitofrontal cortex* (pp. 285–306). Oxford: Oxford University Press.

Brandt, J., & van Gorp, W. G. (2006). Functional ('psychogenic') amnesia. *Seminars in Neurology, 26*, 331–340.

Breitenstein, C., Daum, I., Ackermann, H., Lütgehetmann, R., & Müller, E. (1996). Erfassung der Emotionswahrnehmung bei zentralnervösen Läsionen und Erkrankungen: Psychometrische Gütekriterien der 'Tübinger Affekt Batterie. *Neurologie and Rehabilitation, 2*, 93–101.

Brickenkamp, R., & Zillmer, E. (1998). *d2 test of attention*. Göttingen: Hogrefe.

Brower, M. C., & Price, B. H. (2001). Neuropsychiatry of frontal lobe dysfunction in violent and criminal behaviour: A critical review. *Journal of Neurology, Neurosurgery and Psychiatry, 71*, 720–726.

Bufkin, J. L., & Luttrell, V. R. (2005). Neuroimaging studies of aggressive and violent behavior: Current findings and implications for criminology and criminal justice. *Trauma, Violence and Abuse, 6*, 176–191.

Bush, G., Luu, P., & Posner, M. I. (2000). Cognitive and emotional influences in anterior cingulate cortex. *Trends in Cognitive Science, 4*, 215–222.

Cabeza, R., & St. Jacques, P. (2007). Functional neuroimaging of autobiographical memory. *Trends in Cognitive Sciences, 11*, 219–227.

Cabeza, R., Prince, S. E., Daselaar, S. M., Greenberg, D. L., Budde, M., Dolcos, F., et al. (2004). Brain activity during episodic retrieval of autobiographical and laboratory events: An fMRI study using a novel photo paradigm. *Journal of Cognitive Neuroscience, 16*, 1583–1594.

Carter, C. S., Braver, T. S., Barch, D. M., Botvinick, M. M., Noll, D., & Cohen, J. D. (1998). Anterior cingulate cortex, error detection, and the online monitoring of performance. *Science, 280*, 747–749.

Cavanna, A. E., & Trimble, M. R. (2006). The precuneus: A review of its functional anatomy and behavioural correlates. *Brain, 129*, 564–583.

Conway, M. A., & Pleydell-Pearce, C. W. (2000). The construction of autobiographical memories in the self-memory system. *Psychological Review, 107*, 261–288.

Conway, M. A., Turk, D. J., Miller, S. L., Logan, J., Nebes, R. D., Cidis Meltzer, C., et al. (1999). A positron emission tomography (PET) study of autobiographical memory retrieval. *Memory, 7*, 679–702.

Corcoran, R., & Frith, C. D. (2003). Autobiographical memory and theory of mind: Evidence of a relationship in schizophrenia. *Psychological Medicine, 33*, 897–905.

Craig, J. S., Hatton, C., Craig, F. B., & Bentall, R. P. (2004). Persecutory beliefs, attributions and theory of mind: Comparison of patients with paranoid delusions, Asperger's syndrome and healthy controls. *Schizophrenia Research, 69*, 29–33.

Davidson, R. J., Putnam, K. M., & Larson, C. L. (2000). Dysfunction in the neural circuitry of emotion regulation – a possible prelude to violence. *Science, 289*, 591–594.

de Decker, A., Hermans, D., Raes, F., & Eelen, P. (2003). Autobiographical memory specificity and trauma in inpatient adolescents. *Journal of Clinical Child and Adolescent Psychology, 32*, 22–31.

Deeley, Q., Daly, E., Surguladze, S., Tunstall, N., Mezey, G., Beer, D., et al. (2006). Facial emotion processing in criminal psychopathy. Preliminary functional magnetic resonance imaging study. *British Journal of Psychiatry, 189*, 533–539.

Ekman, P., & Friesen, W. V. (1975). *Pictures of facial affect*. Palo Alto, CA: Consulting Psychologists Press.

Fast, K., Fujiwara, E., & Markowitsch, H. J. (in preparation). *FFT – Famous Faces Test*. Göttingen: Hogrefe.

Fast, K., & Markowitsch, H. J. (in press). Die Erfassung von Alt- und Neugedächtnisleistungen mit dem Autobiographischen Gedächtnis Inventar (AGI). *Zeitschrift für Neuropsychologie*.

Fink, G. R. (2003). In search of one's own past: The neural bases of autobiographical memories. *Brain, 126*, 1509–1510.

Fink, G. R., Markowitsch, H. J., Reinkemeier, M., Bruckbauer, T., Kessler, J., & Heiss, W. D. (1996). Cerebral representation of one's own past: Neural networks involved in autobiographical memory. *Journal of Neuroscience, 16*, 4275–4282.

Fletcher, P. C., Frith, C. D., Baker, S. C., Shallice, T., Frackowiak, R. S., & Dolan, R. J. (1995). The mind's eye – precuneus activation in memory-related imagery. *NeuroImage, 2*, 195–200.

Fletcher, P. C., Shallice, T., Frith, C. D., Frackowiak, R. S., & Dolan, R. J. (1996). Brain activity during memory retrieval. The influence of imagery and semantic cueing. *Brain, 119*, 1587–1596.

Frith, C. D. (2002). Attention to action and awareness of other minds. *Consciousness and cognition, 11*, 481–487.

Frith, C. D. (2007). The social brain? *Philosophical transactions of the Royal Society of London. Series B, Biological sciences, 362*, 671–678.

Fujiwara, E., Brand, M., Kracht, L. W., Kessler, J., Diebel, A., Netz, J., et al. (in press). Functional retrograde amnesia: A multiple case study. *Cortex*.

Gehring, W. J., & Knight, R. T. (2000). Prefrontal-cingulate interactions in action monitoring. *Nature Neuroscience, 3*, 516–520.

Greicius, M. D., Krasnow, B., Reiss, A. L., & Menon, V. (2003). Functional connectivity in the resting brain: A network analysis of the default mode hypothesis. *Proceedings of the National Academy of Sciences of the United States of America, 100*, 253–258.

Hampson, M., Driesen, N. R., Skudlarski, P., Gore, J. C., & Constable, R. T. (2006). Brain connectivity related to working memory performance. *Journal of Neuroscience, 26*, 13338–13343.

Happé, F. G. (1994). An advanced test of theory of mind: Understanding of story characters' thoughts and feelings by able autistic, mentally handicapped, and normal children and adults. *Journal of Autism and Developmental Disorders, 24*, 129–154.

Hare, R. D. (2003). *The Psychopathy Checklist-Revised.* Toronto: Multi-Health Systems.

Harrington, L., Langdon, R., Siegert, R. J., & McClure, J. (2005a). Schizophrenia, theory of mind, and persecutory delusions. *Cognitive Neuropsychiatry, 10*, 87–104.

Harrington, L., Siegert, R. J., & McClure, J. (2005b). Theory of mind in schizophrenia: A critical review. *Cognitive Neuropsychiatry, 10*, 249–286.

Härting, C., Markowitsch, H. J., Neufeld, H., Calabrese, P., & Deisinger, K. (2000). *Wechsler Gedächtnis Test – Revidierte Fassung. Deutsche Adaptation der revidierten Fassung der Wechsler-Memory-Scale.* Göttingen: Hogrefe.

Hester, R., Fassbender, C., & Garavan, H. (2004). Individual differences in error processing: A review and reanalysis of three event-related fMRI studies using the GO/NOGO task. *Cerebral Cortex, 14*, 986–994.

Horn, W. (1983). *Leistungsprüfsystem.* Göttingen: Hogrefe.

Johnson, S. C., Baxter, L. C., Wilder, L. S., Pipe, J. G., Heiserman, J. E., & Prigatano, G. P. (2002). Neural correlates of self-reflection. *Brain, 125*, 1808–1814.

Kalbe, E., Kessler, J., Calabrese, P., Smith, R., Passmore, A. P., Brand, M., et al. (2004). DemTect: A new, sensitive cognitive screening test to support the diagnosis of mild cognitive impairment and early dementia. *International Journal of Geriatric Psychiatry, 19*, 136–143.

Kaney, S., Bowen-Jones, K., & Bentall, R. P. (1999). Persecutory delusions and autobiographical memory. *The British Journal of Clinical Psychology, 38*, 97–102.

Keedwell, P. A., Andrew, C., Williams, S. C., Brammer, M. J., & Phillips, M. L. (2005). A double dissociation of ventromedial prefrontal cortical responses to sad and happy stimuli in depressed and healthy individuals. *Biological Psychiatry, 58*, 495–503.

Kiehl, K. A., Smith, A. M., Hare, R. D., Mendrek, A., Forster, B. B., Brink, J., et al. (2001). Limbic abnormalities in affective processing by criminal psychopaths as revealed by functional magnetic resonance imaging. *Biological Psychiatry, 50*, 677–684.

Kjaer, T. W., Nowak, M., & Lou, H. C. (2002). Reflective self-awareness and conscious states: PET evidence for a common midline parietofrontal core. *NeuroImage, 17*, 1080–1086.

Kondo, H., Osaka, N., & Osaka, M. (2004). Cooperation of the anterior cingulate cortex and dorsolateral prefrontal cortex for attention shifting. *NeuroImage, 23*, 670–679.

Kopelman, M. D. (2000). Focal retrograde amnesia and the attribution of causality: An exceptionally critical review. *Cognitive Neuropsychology, 17*, 585–621.

Kopelman, M. D., Wilson, B. A., & Baddeley, A. D. (1989). The autobiographical memory interview: A new assessment of autobiographical and personal semantic memory in amnesic patients. *Journal of Clinical and Experimental Neuropsychology, 11*, 724–744.

Kritchevsky, M., Chang, J., & Squire, L. R. (2004). Functional amnesia: Clinical description and neuropsychological profile of 10 cases. *Learning and Memory, 11*, 213–226.

Langleben, D. D., Loughead, J. W., Bilker, W. B., Ruparel, K., Childress, A. R., Busch, S. I., et al. (2005). Telling truth from lie in individual subjects with fast event-related fMRI. *Human Brain Mapping, 26*, 262–272.

Lee, L., Harkness, K. L., Sabbagh, M. A., & Jacobson, J. A. (2005). Mental state decoding abilities in clinical depression. *Journal of Affective Disorders, 86*, 247–258.

Leplow, B., Blunck, U., Schulze, K., & Ferstl, R. (1993). Der Kieler Altgedächtnistest: Neuentwicklung eines deutschsprachigen Famous Event-Tests zur Erfassung des Altgedächtnisses. *Diagnostica, 39*, 240–256.

Leplow, B., Dierks, C., Merten, T., & Hänsgen, K. (1997). Probleme des Geltungsbereiches deutschsprachiger Altgedächtnistests. *Zeitschrift für Neuropsychologie, 8*, 137–144.

Levine, B. (2004). Autobiographical memory and the self in time: Brain lesion effects, functional neuroanatomy, and lifespan development. *Brain and Cognition, 55*, 54–68.

Lie, C. H., Specht, K., Marshall, J. C., & Fink, G. R. (2006). Using fMRI to decompose the neural processes underlying the Wisconsin Card Sorting Test. *NeuroImage, 30*, 1038–1049.

Liotti, M., Mayberg, H. S., McGinnis, S., Brannan, S. L., & Jerabek, P. (2002). Unmasking disease-specific cerebral blood flow abnormalities: Mood challenge in patients with remitted unipolar depression. *American Journal of Psychiatry, 159*, 1830–1840.

Lou, H. C., Luber, B., Crupain, M., Keenan, J. P., Nowak, M., Kjaer, T. W., et al. (2004). Parietal cortex and representation of the mental self. *Proceedings of the National Academy of Sciences of the United States of America, 101*, 6827–6832.

Luu, P., Flaisch, T., & Tucker, D. M. (2000). Medial frontal cortex in action monitoring. *Journal of Neuroscience, 20*, 464–469.

Maddock, R. J., Garrett, A. S., & Buonocore, M. H. (2001). Remembering familiar people: The posterior cingulate cortex and autobiographical memory retrieval. *Neuroscience, 104*, 667–676.

Maguire, E. A., & Frith, C. D. (2003). Lateral asymmetry in the hippocampal response to the remoteness of autobiographical memories. *Journal of Neuroscience, 23*, 5302–5307.

Maguire, E. A., Henson, R. N., Mummery, C. J., & Frith, C. D. (2001). Activity in prefrontal cortex, not hippocampus, varies parametrically with the increasing remoteness of memories. *NeuroReport, 12*, 441–444.

Markowitsch, H. J. (1995). Which brain regions are critically involved in retrieval of old episodic memory? *Brain Research Brain Research Reviews, 21*, 117–127.

Markowitsch, H. J. (1996a). Organic and psychogenic amnesia: Two sides of the same coin? *Neurocase, 2*, 357–371.

Markowitsch, H. J. (1996b). Retrograde amnesia: Similarities between organic and psychogenic forms. *Neurology, Psychiatry and Brain Research, 4*, 1–8.

Markowitsch, H. J. (1999a). Functional neuroimaging correlates of functional amnesia. *Memory, 7*, 561–583.

Markowitsch, H. J. (1999b). Stress-related memory disorders. In L. G. Nilsson, & H. J. Markowitsch (Eds), *Cognitive neuroscience of memory* (pp. 193–211). Göttingen: Hogrefe.

Markowitsch, H. J. (2000). The anatomical bases of memory. In M. S. Gazzaniga (Ed.), *The new cognitive neurosciences* (pp. 781–795). Cambridge, MA: The MIT Press.

Markowitsch, H. J. (2003a). Memory: Disturbances and therapy. In T. Brandt, L. Caplan, J. Dichgans, H. C. Diener, & C. Kennard (Eds), *Neurological disorders; course and treatment* (pp. 287–302). San Diego, CA: Academic Press.

Markowitsch, H. J. (2003b). Psychogenic amnesia. *NeuroImage, 20*, S132–S138.

Markowitsch, H. J., & Siefer, W. (2007). *Tatort Gehirn. Auf der Suche nach dem Ursprung des Verbrechens.* München: Beck.

Markowitsch, H. J, Kessler, J., Van der Ven, C., Weber-Luxenburger, G., & Heiss, W. D. (1998). Psychic trauma causing grossly reduced brain metabolism and cognitive deterioration. *Neuropsychologia, 36*, 77–82.

Markowitsch, H. J., Thiel, A., Reinkemeier, M., Kessler, J., Koyuncu, A., & Heiss, W. D. (2000). Right amygdala and temporofrontal activation during autobiographic, but not during fictitious memory retrieval. *Behavioural Neurology, 12*, 181–90.

Markowitsch, H. J., Vandekerckhove, M. M., Lanfermann, H., & Russ, M. O. (2003). Engagement of lateral and medial prefrontal areas in the ecphory of sad and happy autobiographical memories. *Cortex, 39*, 643–665.

Müller, J. L., Sommer, M., Wagner, V., Lange, K., Taschler, H., & Roder, C. H., et al. (2003). Abnormalities in emotion processing within cortical and subcortical regions in criminal psychopaths: Evidence from a functional magnetic resonance imaging study using pictures with emotional content. *Biological Psychiatry, 54*, 152–162.

Otsuka, Y., Osaka, N., Morishita, M., Kondo, H., & Osaka, M. (2006). Decreased activation of anterior cingulate cortex in the working memory of the elderly. *Neuroreport, 17*, 1479–1482.

Piefke, M., & Fink, G. R. (2005). Recollections of one's own past: The effects of aging and gender on the neural mechanisms of episodic autobiographical memory. *Anatomy and Embryology, 210*, 497–512.

Piefke, M., Weiss, P. H., Zilles, K., Markowitsch, H. J., & Fink, G. R. (2003). Differential remoteness and emotional tone modulate the neural correlates of autobiographical memory. *Brain, 126*, 650–668.

Piefke, M., Weiss, P. H., Markowitsch, H. J., & Fink, G. R. (2005). Gender differences in the functional neuroanatomy of emotional episodic autobiographical memory. *Human Brain Mapping, 24*, 313–324.

Raine, A., Stoddard, J., Bihrle, S., & Buchsbaum, M. (1998). Prefrontal glucose in murderers lacking psychosocial deprivation. *Neuropsychiatry, Neuropsychology, and Behavioral Neurology, 11*, 1–7.

Raine, A., Lencz, T., Bihrle, S., LaCasse, L., & Colletti, P. (2000). Reduced prefrontal gray matter volume and reduced autonomic activity in antisocial personality disorder. *Archives of General Psychiatry, 57*, 119–127.

Raine, A., Moffitt, T. E., Caspi, A., Loeber, R., Stouthamer-Loeber, M., & Lynam, D. (2005). Neurocognitive impairments in boys on the life-course persistent antisocial path. *Journal of Abnormal Psychology, 114*, 38–49.

Reinhold, N., Kühnel, S., Brand, M., & Markowitsch, H. J. (2006). Functional brain imaging in memory and memory disorders. *Current Medical Imaging Reviews, 2*, 35–57.

Riutort, M., Cuervo, C., Danion, J. M., Peretti, C. S., & Salame, P. (2003). Reduced levels of specific autobiographical memories in schizophrenia. *Psychiatry Research, 117*, 35–45.

Ryan, L., Nadel, L., Keil, K., Putnam, K., Schnyer, D., Trouard, T., et al. (2001). Hippocampal complex and retrieval of recent and very remote autobiographical memories: Evidence from functional magnetic resonance imaging in neurologically intact people. *Hippocampus, 11*, 707–714.

Schore, A. N. (2002). Dysregulation of the right brain: A fundamental mechanism of traumatic attachment and the psychopathogenesis of posttraumatic stress disorder. *Australian and New Zealand Journal of Psychiatry, 36*, 9–30.

Sellal, F., Manning, L., Seegmuller, C., Scheiber, C., & Schoenfelder, F. (2002). Pure retrograde amnesia following a mild head trauma: A neuropsychological and metabolic study. *Cortex, 38*, 499–509.

Serra, L., Fadda, L., Buccione, I., Caltagirone, C., & Carlesimo, G. A. (2007). Psychogenic and organic amnesia: A multidimensional assessment of clinical, neuroradiological, neuropsychological and psychopathological features. *Behavioural Neurology, 18*, 53–64.

Shima, K., & Tanji, J. (1998). Role for cingulate motor area cells in voluntary movement selection based on reward. *Science, 282*, 1335–1338.

Sonntag, P., Gokalsing, E., Olivier, C., Robert, P., Burglen, F., Kauffmann-Muller, F., et al. (2003). Impaired strategic regulation of contents of conscious awareness in schizophrenia. *Consciousness and Cognition, 12*, 190–200.

Spence, S. A., Farrow, T. F., Herford, A. E., Wilkinson, I. D., Zheng, Y., & Woodruff, P. W. (2001). Behavioural and functional anatomical correlates of deception in humans. *Neuroreport, 12*, 2849–2853.

Strauss, E., Sherman, E. M. S., & Spreen, O. (2006). *A compendium of neuropsychological tests. Administration, norms, and commentary.* Oxford: Oxford University Press.

Sugiura, M., Shah, N. J., Zilles, K., & Fink, G. R. (2005). Cortical representations of personally familiar objects and places: Functional organization of the human posterior cingulate cortex. *Journal of Cognitive Neuroscience, 17*, 183–198.

Svoboda, E., McKinnon, M. C., & Levine, B. (2006). The functional neuroanatomy of autobiographical memory: A meta-analysis. *Neuropsychologia, 44*, 2189–2208.

Thiel, A., Herholz, K., Koyuncu, A., Ghaemi, M., Kracht, L. W., Habedank, B., et al. (2001). Plasticity of language networks in patients with brain tumors: A positron emission tomography activation study. *Annals of Neurology, 50*, 620–629.

Tulving, E. (2002). Episodic memory: From mind to brain. *Annual Reviews of Psychology, 53*, 1–25.

Tulving, E. (2005). Episodic memory and autonoesis: Uniquely human? In H. Terrace, & J. Metcalfe (Eds), *The missing link in cognition: Evolution of self-knowing consciousness* (pp. 3–56). New York: Oxford University Press.

Uddin, L. Q., Kaplan, J. T., Molnar-Szakacs, I., Zaidel, E., & Iacoboni, M. (2005). Self-face recognition activates a frontoparietal 'mirror' network in the right hemisphere: An event-related fMRI study. *NeuroImage, 25*, 926–935.

van Veen, V., & Carter, C. S. (2006). Error detection, correction, and prevention in the brain: A brief review of data and theories. *Clinical EEG and Neuroscience, 37*, 330–335.

Van Vreeswijk, M. F., & De Wilde, E. J. (2004). Autobiographical memory specificity, psychopathology, depressed mood and the use of the Autobiographical Memory Test: A meta-analysis. *Behavioural Research and Therapy, 42*, 731–743.

Vandekerckhove, M. M. P., Markowitsch, H. J., Mertens, M., & Woermann, F. G. (2005). Bi-hemispheric engagement in the retrieval of autobiographical episodes. *Behavioural Neurology, 16*, 203–210.

Viard, A., Piolino, P., Desgranges, B., Chetelat, G., Lebreton, K., Landeau, B., et al. (2007). Hippocampal activation for autobiographical memories over the entire lifetime in healthy aged subjects: An fMRI study. *Cerebral Cortex*, doi:10.1093/cercor/bhl153.

Vogeley, K., May, M., Ritzl, A., Falkai, P., Zilles, K., & Fink, G. R. (2004). Neural correlates of first-person perspective as one constituent of human self-consciousness. *Journal of Cognitive Neuroscience, 16*, 817–827.

von Cramon, D., Markowitsch, H. J., & Schuri, U. (1993). The possible contribution of the septal region to memory. *Neuropsychologia, 31*, 1159–1180.

Williams, J. M. G. (1996). Depression and the specificity of autobiographical memory. In D. C. Rubin (Ed.), *Remembering our past* (pp. 244–267). New York, NY: Cambridge University Press.

Williams, J. M., Barnhofer, T., Crane, C., Herman, D., Raes, F., Watkins, E., et al. (2007). Autobiographical memory specificity and emotional disorder. *Psychological Bulletin, 133*, 122–148.

Yeung, N., Holroyd, C. B., & Cohen, J. D. (2005). ERP correlates of feedback and reward processing in the presence and absence of response choice. *Cerebral Cortex, 15*, 535–544.

NEUROCASE
2008, 14 (1), 29–43

Kindled non-convulsive behavioral seizures, analogous to primates. A 24th case of 'limbic psychotic trigger reaction': Bizarre parental infanticide — might nonvoluntariness during LPTR become objectified by primate model?

Anneliese A. Pontius

Harvard Medical School (ret.), Boston, MA, USA

Limbic psychotic trigger reaction (LPTR) includes paroxysmal, out-of-character, motiveless, unplanned felonies (or similarly bizarre social misbehavior), all committed during flat affect, autonomic arousal and a fleeting *de novo* psychosis. A transient limbic hyperactivation is implicated that impairs prefrontal monitoring (judgment, planning, intent, volition, emotional participation) but preserves memory for the acts. It is hypothesized that LPTR implicates an atavistic regression to a limbic 'paleo-consciousness', exemplified by a 24th patient (parental infanticide), presented herein. He had closed head injury and borderline abnormal posterior brain pathology (EEG/CT), which might have contributed to his unusually numerous visual hallucinations.

Keywords: Partial limbic seizures; Seizure kindling; Primate model; Limbic psychotic trigger reaction; Social loners; Paleo-consciousness; Nonvolitional; Motiveless criminal acts; Social misbehavior; Memory retention.

INTRODUCTION

This study provides an explanation for hitherto unexplainable bizarre felonious acts by the hypothesis of a lowered threshold for neurophysiological seizure kindling in certain individuals as described in Tables 1 and 2. This hypothesis is capable of uniting seemingly unrelated symptoms and signs into a neurophysiologically meaningful pattern, and clinical syndrome of partial seizures, called limbic psychotic trigger reaction (LPTR) (Cromie, 1996; Landau, 1996; LoPiccolo, 1996; Pontius, 1981, 1984, 1986, 1987, 1990, 1993a, 1993b, 1993c, 1995, 1996a, 1996b, 1997, 1999, 2000, 2001a, 2001b,

2002a, 2002b, 2003, 2004, 2005; Pontius & LeMay, 2003; Pontius & Wieser, 2004; Shostakovitch & Leonova, 2004). Outstanding indications for a seizure syndrome are particularly its paroxysmal course and its emergence in three seizure-like phases (aura, ictus, post-ictus; see Table 1). Furthermore, a specifically kindled kind of seizure is implicated by its elicitation upon an encounter with an individualized trigger stimulus. Such a stimulus represents directly or symbolically similar intermittent past exposures to merely mild or moderate stressful experiences (herein worries about mild to moderate child neglect). Together, this specific pattern of symptoms and signs is strongly

A brief version of this article has been presented by invitation of the Russian Academy of Sciences at 3rd International Interdisciplinary Conference 'Neuroscience for Medicine and Psychology' in Sudak, Crimea, Ukraine, 2007.

Address correspondence to Anneliese A. Pontius, MD, Associate Clinical Professor (ret.), Waldschmidt St 6, 60316 Frankfurt/M, Germany (E-mail: anneliese_pontius@hms.harvard.edu).

http://www.psypress.com/neurocase   DOI: 10.1080/13554790801992750

TABLE 1
Symptomatology of limbic psychotic trigger reaction

(1) A paroxysmal sequence of automatized behaviors, emerging in three seizure-like phases (aura, ictus, post-ictus).
(2) Following a chance encounter with an individualized trigger stimulus (of any modality or symbolic) that revives the memory of intermittent, merely mild to moderate stressful experiences (analogous to experimental seizure kindling).
(3) Aura-like symptoms (ca. 5 min) (e.g., a consternating 'cognitive mismatch', and/or autonomic arousal with epigastric sensations, nausea, vertigo, 'ice cold' sensations, profuse sweating, 'tingling', tinnitus, diarrhea, urinary incontinence, erection, or ejaculation.
(4) First time fleeting psychotic symptoms (hallucinations of any modality, frequently visual, formed or unformed, and/or delusions, frequently of grandeur). If previous psychotic symptoms had been present, they take on a different psychotic content.
(5) Out-of-character non-voluntary acts.
(6) Flat affect around the time of the act.
(7) Lack of emotional motivation.
(8) Lack of preplanning of the act or of its concealment.
(9) Virtually fully retained memory of the act (compatible with an implicated atavistic regression to limbic 'paleo-consciousness').
(10) Indications of transient pre-frontal dysfunctioning around the time of the act as part of an implicated temporary fronto-limbic dysbalance, where limbic seizure activation might transiently overwhelm prefrontal monitoring.
(11) Post-ictal lingering (for some hours) of prefrontal dysfunction, e.g., inefficient 'stupid' behavior and/or lingering distortions of space or time.
(12) Mostly self-confession of the act.
(13) Subjective assumption of responsibility with remorse (and/or suicidal ideation – unless there is an underlying schizophrenia with habitually flat affect).

LPTR consists of a neurophysiologically linked pattern of 13 interrelated symptoms that may vary by degree, but not in essence. The symptomatology is determined by 16 inclusion and 13 exclusion criteria (Pontius & Wieser, 2004, Table 1).

TABLE 2
Analogies between experimental seizure kindling in primates vs. naturalistic seizure kindling implicated in limbic psychotic trigger reaction

Kindling of primates	*Limbic psychotic trigger reaction*
No brain lesions required (Post & Kopanda, 1975; Wada, 1978, 1990)	Same, although lesions might facilitate kindling
Nonconvulsive behavioral seizures	Same
Elicited by intermittent exposure to subthreshold stimuli (electrical, chemical and/or experiential)	By intermittent exposure to mild/moderate experiential stimuli and/or chemical (Pontius, 2001b)
In mammals memories of emotional stimuli alone had the same physiological effects as actual stimuli (Cromie, 2002; Kandel, 2001)	Or only to memories of stimuli
Unusual out-of-species-behavior with indications of visual hallucinations (e.g., grabbing in the air)	Bizarre, out-of-character behavior with psychotic symptoms: visual and/or other hallucinations and/or delusions (often of grandeur)

Note: The hypothesis of seizure kindling in LPTR had evolved through the scientific method of retroductive inference (Harre, 1970; Kuhn, 1977), elaborated e.g., by theoretical physicist Hanson (1965) and by Nobel economist Hayek (1964), as previously detailed (Pontius, 1995): (1) observation of consistently repeated naturalistic, hitherto unexplainable phenomena (here symptoms and signs of LPTR, Table 1). (2) A hypothesis is proposed, which, if it were true, could explain the phenomena (here the hypothesis of seizure kindling, a neuro-physiological mechanism). (3) The 'puzzle solving' phase (Kuhn, 1977) includes the essential process of 'choosing a generative mechanism for the phenomena within a plausible model, a mechanism capable of uniting and explaining the heretofore unexplainable phenomena' (Harre, 1970) (here choosing the neuro-physiological mechanism of seizure kindling, analogous to non-convulsive behavioral seizure kindling in primates). (4) 'Testing the plausible model is an ultimate, often belated process' (Kuhn, 1977). It includes the weighty criteria of (a) 'accuracy' (see Primate Model); (b) scope or meaningfulness; (c) 'fruitfulness' (see 'Research on LPTR'); and (d) 'consistency with a corpus of accepted scientific knowledge' (previously detailed, Pontius, 1995, and below). Thus, LPTR symptomatology is consistent with the generally accepted characteristic of simple seizures (with preservation of consciousness), during which EEG abnormalities are rare (e.g., occurring only in 15% of ongoing recordings of observed simple seizures, Devinsky, Kelley, Porter, & Theodore, 1988). Moreover, aside from LPTR's primate model, a certain consistency also pertains to certain human experimentation with direct electrical stimulation of brain implants in pre-surgery patients (Gloor et al., 1982; detailed by Pontius, 1997); and Wieser (1983, 1987), detailed by Pontius and Wieser (2004).

analogous to experimental kindling, particularly of non-convulsive behavioral seizures in primates with indications of visual hallucinations (Table 2).

Experimental seizure kindling requires no brain lesions and so far none have been detected afterwards (Cain, 1992). Instead, kindled seizures are evoked by intermittent exposure to external and merely subthreshold stimuli (electrical, chemical or experiential), whereby each stimulus in itself is innocuous (Adamec, 1987; Corcoran & Solomon, 2005; Ebert & Loescher, 1999; Goddard, 1967; Goddard & McIntyre, 1986; Krishnamoorthy & Trimble, 1998; Racine, 1980; Racine & Burnham, 1982).

In lower mammals kindling elicits convulsions, but in primates it elicits prevailingly non-convulsive behavioral seizures (NCBS) with indications of visual hallucinations (Post & Kopanda, 1975; Wada, 1978, 1990).

To the few human cases of inadvertent kindling reported in the literature (Heath, Monroe, & Mickle, 1955; Morrell, 1985; Sramka, Sedlak, & Nadvornik, 1984), an additional 24 LPTR patients can be added. In these cases the kindling phenomena are expressed in bizarre, unusual and sometimes criminal patterns of behavior.

Such patients' bizarre pattern of behavior does not fit any 'crime profiles', nor any traditional diagnoses (Garland, 2004). So far, among 24 LPTR patients, there have been 19 'murderers', four 'fire setters', one 'bank robber'.

All LPTR patients have repeatedly and consistently presented with 13 symptoms and signs (Table 1), determined by 16 inclusion and 13 exclusion criteria (Pontius, 1997; Pontius & Wieser, 2004, Table 1).

LPTR may, however, not be limited to felonies which have attracted forensic attention. In addition there may exist many more 'merely' social misbehaviors, undetected and untreated as 'sleeper' cases of LPTR. Proust (1913) may have presented such a 'case' in his detailed description of M. Swann, who officially proposed marriage to Odette, known to him as a 'despicable cocotte' upon hearing a brief piece of music, reviving individualized memories (Pontius, 1993b; Pontius & Wieser, 2004).

An additional neuro-physiological factor relevant to LPTR is suggested by mammalian experiments demonstrating that actual emotional stimuli can be replaced by their memories alone producing similar physiological effects (Cromie, 2002; Kandel, 2001).

In combination such an effect of memories in mammals and kindling in primates are congruent with the finding that all 24 LPTR patients had been socially deprived (Fuchs, 2005) loners (schizoid personalities and/or presenting alexithymia, as herein). In part, loneliness might be associated with such patients' tendency not to share or compare their merely mild to moderate stresses with others, not laying them to rest, but tending to ruminate on them intermittently in memory. Such a scenario is analogous to that required for seizure kindling

(Table 1). After several repetitions, kindling in LPTR appears to be finally triggered by a chance encounter with an individualized external stimulus (or with several similar stimuli within specific context). This final triggering stimulus, unbeknownst to the patient probably represents the last stimulus in a row of intermittently preceding similar stimuli required for kindling and accordingly all such stimuli involve merely mild to moderate stresses (herein worries about child neglect were symbolized by the sight of an earring for a pierced ear).

So far, no consistent abnormalities on objective brain tests have been detected, which is consistent with experimental seizure kindling requiring no brain lesion. However, a damaged brain might be more susceptible to seizure kindling. So far, half of the 24 LPTR patients have had (frequently overlooked) closed head injuries (including the new patient reported here) (Levin et al., 1989), and half have had some abnormality on EEG, CT scan or MRI at some time during their lives. The new patient's EEG and CT scan revealed borderline pathology in the posterior brain region. This might have been an incidental finding, common within the naturalistic setting, or it might have contributed to the patient's unusually extensive visual illusions and hallucinations, particularly regarding the human face.

An overriding conceptualization of the LPTR symptomatology suggests a seizure-based temporary fronto-limbic imbalance, whereby a limbic seizure activation would briefly overwhelm the reciprocally related (Nauta, 1971) prefrontal monitoring. This imbalance would eventuate in a brief altered state of consciousness without impairment of memory for the bizarre acts, called limbic 'paleoconsciousness' (congruent with MacLean's generalistic conceptualizing, 1990, 1992). This state could be associated with the out-of-character felonious acts that implicate severely regressed prefrontal functions of motivation, judgment, intent, planning, volition (Halleck, 1992), emotional involvement (Damasio, 1999) and an impaired sense of 'self' (Clore, 1992; Pontius & Wieser, 2004, section 3.4).

SPECIFIC CONSISTENCY OF LPTR WITH ESTABLISHED KNOWLEDGE BASED ON EXPERIMENTALLY EVOKED BRAIN IMPLANTS IN PRE-SURGERY PATIENTS

Intriguingly, certain LPTR symptoms (including automatism and 'mental or psychic phenomena')

TABLE 3
Two reverse homicidal syndromes: Fried's 'Syndrome E' vs. LPTR show reverse
behavior and correspondingly implicate reverse neuropathology

---

Stark contrast between two hypothesized reverse syndromes: 'Syndrome E' and 'LPTR'

• LPTR presents paroxysmal, motiveless, unplanned, non-voluntary acts, committed without emotional involvement by a single patient committing homicide upon a chance encounter with an individualized trigger stimulus, apparently kindling a limbic seizure (Table 2).

• 'Syndrome E' occurs over long periods of time, without indications of seizures. Obsessive ideation prevails within a group-related ideology directed against pre-selected defenseless victims. Rapid desensitization occurs after initial 'elation', later followed by diminished affective reactivity. There is no evidence of psychosis, only of autonomic arousal. The acts show a general environmental dependence and group contagion.

Implicated neuropathology suggests fronto-limbic dysbalance in both syndromes, but in reverse

• In 'Syndrome E' there is 'deficient interaction between prefrontal cortices and amygdala', as Fried (1997) put it: 'The amygdala is tonically inhibited by prefrontal activation'.

• In LPTR fleeting seizure-related initial limbic (amygdalar) hyperactivation is associated with transient secondary prefrontal dysfunction, given that both systems are reciprocally interrelated (Nauta, 1971).

Implication of particular prefrontal regions in reverse in 'Syndrome E' vs. LPTR

(a) Dorsolateral prefrontal cortex
  • In 'Syndrome E' it is relatively unaffected, leaving planning and problem solving intact.
  • In LPTR it is transiently impaired, resulting in lack of planning and deficient problem solving.

(b) Orbitofrontal cortex
  • In 'Syndrome E' it is hyperactive, reflected by obsessive ideation and repetitive acts.
  • In LPTR it is normally active: typically only a single act occurs at a time.

(c) Medial prefrontal cortex
  • In 'Syndrome E' it is hyperactive, expressed in high motivation and elation.
  • In LPTR the reverse may be implicated, because acts are motiveless and consistently associated with flat affect.

(d) Ventromedial prefrontal cortex
  • In 'Syndrome E' it is affected, not generating 'emotions appropriate to the images conjured by certain acts or situational stimuli' (Fried, 1997).
  • In LPTR this region might be transiently affected, possibly contributing to the exaggerated albeit fleeting impact of individualized trigger stimuli within a specific situational context, reviving memories of merely mild to moderate stressful experiences.

Extremely opposite clusters of behavior and correspondingly reverse implication of neuropathology pertain to two opposite homicidal syndromes: LPTR (Table 1) vs. 'Syndrome E', hypothesized by neuroscientist Fried (1997) in regard to ethnic murderers world wide.

---

are similar to those evoked by direct electrical stimulation of brain implants in the mesotemperobasal brain region of pre-surgery patients (Wieser, 1983, 1989). As can be expected, however, the electrically elicited symptoms were not as rich in scope as those occurring within the naturalistic setting of LPTR (Pontius & Wieser, 2004) Under similar experimental conditions, Gloor et al. (1982) too had elicited 'psychic phenomena' analogous to LPTR by stimulating such patients' amygdala, as previously indicated (Pontius, 1997, Table 4).

INDIRECT SUPPORT FOR THE SYMPTOMATOLOGY OF LPTR AND ITS NEUROPATHOLOGY PROVIDED BY THE REVERSE 'MIRROR-IMAGE' OF FRIED'S 'SYNDROME E' OF ETHNIC MURDERERS

An independently hypothesized 'Syndrome E' (Fried, 1997) in ethnic murderers world-wide, presents a surprising mirror-image in reverse to the proposed LPTR. The reversal both in regard to behavior and to the detailed neuropathology (implicated by neuroscientist Fried), provides serendipitous mutual support for each one of these two syndromes, as previously detailed (Pontius, 2000, Table 1), and briefly summarized in Table 3.

BRIEF REVIEW OF ONE PREVIOUS EXCEPTIONAL CASE OF LPTR (HOMICIDE OF CLOSELY RELATED PERSONS: GRAND-CHILDREN)

As will be detailed later on, kindling (requiring merely intermittent exposure to stimuli) is considered to be the reverse to 'tolerance' (Post & Kopanda, 1975). Tolerance tends to develop between persons sharing living quarters. Accordingly, among the 24 cases of LPTR there had only been three involving closely living persons: a brother (Pontius, 1981, 1984); an infant daughter (Pontius, 1990, 2003a, 2003b); and the following case of grand-parental fillicide (Pontius, 2005). Intriguingly this patient's psychotic symptoms involved visual hallucinations involving the human face, as in the new patient.

A devoted grandfather, a social loner in his 60s, had been a physically healthy retiree. All his laboratory findings, including a routine scalp EEG had been within normal limits. Outstanding among his otherwise scarce childhood memories had been his father's introduction to deer hunting during his teens as 'a manly thing to do'. Although he had previously been an ambivalent deer hunter, he had begun to hunt again during his recent retirement. He had been devoted to his two preteen grandchildren (a boy and a girl). One day upon returning from a successful deer hunt, he suddenly suffered unusual stomach aches and began to puzzle over the changing 'strange faces' of his grandchildren looking at him while standing with their backs to the fire place with hunting photos on its mantelpiece. The photos and the preteens' faces had suddenly revived his memories of killed deer. Suddenly, 'without any reason', he grabbed his shot gun standing still next to him (reminiscent of 'ulilization behavior' (L'hermitte, 1986), and fatally shot his 'changed' grandchildren 'without any feeling', motive, intent or volition. As he remembered his deeds with horror and guilt, he seriously attempted to shoot himself too.

DIFFERENTIAL DIAGNOSES TO THE PROPOSED LPTR

This brief survey of previously discussed differential DSM-IV diagnoses of LPTR (Pontius, 1981–2005; Pontius & Wieser, 2004) will again underscore that LPTR is not explainable by any of the traditional diagnoses beyond vague phenomenological ones without heuristic value, such as 'acute psychosis'. All of the following diagnoses do not meet the strict criteria of LPTR (Pontius & Wieser, 2004, Table 1).

a. Temporal lobe epilepsy (TLE) (Bear, 1987; Geschwind, 1984; Pontius, 2001; Pontius & LeMay, 2003), another subtype of partial seizures shows a certain overlap with LPTR but begins with intense emotions (dread, elation) and is typically associated with clouding of consciousness with amnesia for the acts. Similar to LPTR, TLE acts are also seemingly volitional acts, e.g., during a dissociate fugue, TLE patients can travel around, buying bus tickets, but do not remember their acts afterwards. The clouding of consciousness prevents TLE patients from presenting a danger to others (Delgado-Escueta et al., 1981). Furthermore, the typical amnesia in TLE limits detailed clinical exploration in contrast to LPTR.

b. Disorders of impulse control, such as out of context intermittent explosive or conduct disorders as well as compulsivity disorder (Kuzma & Black, 2004), recur intermittently over years, frequently with childhood onset. Typically such aggressive or hostile personalities do not act out-of-character and experience strong emotions, especially rage (in stark contrast to the flat affect characterizing LPTR).

c. Schizophrenia (Pontius, 2004) is typically associated with cognitive and subtle but characteristic motor system disturbances and with poor general living and coping skills. Schizophrenia (as well as the other major psychotic syndromes) lasts longer than the paroxysmal LPTR. Furthermore, visual hallucinations are rare in schizophrenia, where auditory ones prevail.

d. Panic attacks (frequently with childhood onset), are characterized by typically longstanding avoidant behavior under the impact of intense conscious feelings of fear (particularly of suffocation or of disgust).

e. Psychopathic or sociopathic personality disorders are of long standing. The acts are compatible with the person's character ('egosyntonic') and are typically planned and motivated by various kinds of gains.

f. Reflex epilepsies (Geschwind & Sherwin, 1967) are elicited by one entire stimulus modality, such as by writing or reading in general, in contrast to an individualized stimulus triggering LPTR.

g. Delusional misidentification syndrome (Aziz, Razik, & Donn, 2005) involves the delusional belief of replacement by doubles or imposters of people, animals or objects. So far there has been no case of LPTR with such symptoms, merely involving certain changed faces or distortions of other features on otherwise correctly identified persons.

h. 'Breakthrough reaction' (a former psychoanalytic diagnosis, no longer listed in DSM) had been hypothesized to occur after formerly repressed instinctual needs suddenly become manifest and acted upon with strong emotions (in contrast to the flat affect during LPTR).

CASE REPORT

A previously non-aggressive, fully employed white engineer in his 30s suddenly suffocated his two beloved infants. His motiveless, non-volitional, unplanned acts occurred during visual hallucinations, without emotional involvement (in distinct contrast to impulsive disorders).

METHODS

Two kinds of methods have been used: (A) a study of available Court reports and (B) new additional Court-approved neuropsychiatric interviews.

(A) Studying reports by previous psychiatrists

(A1)

Two previous psychiatric examiners had not found any axis 1 diagnosis under DSM-IV, nor under ICD-10. They had merely suggested 'alexithymia' and 'dissociate disorder' without etiological specification. The previous examiners concluded accordingly that the patient's 'personality traits were not grave enough to lead to serious abnormalities'; therefore, they did not meet the requirements for 'not guilty by reason of insanity'. Wisely, however, both psychiatrists had considered it advisable 'to gain additional future insights about this patient' and the EEG expert had advised: 'The suspected brain dysfunction under provocation needs further clinical investigation'.

Tellingly and unwittingly the psychiatrists felt that their diagnostic impression was especially supported by the fact that the patient's memory for his acts was essentially intact, and that he lacked emotional involvement – these symptoms, however, are precisely essential ones in the diagnosis of LPTR, to be proposed herein.

(A2) Previous reports on 10 neuropsychological test performances

a. The Hamburg Wechsler Intelligence Test (Total IQ 124; Verbal IQ 120; Performance IQ 114). The examiner noted, however, that the patient had not been fully concentrated so that his IQ could be potentially higher.
b. Examination of general ability to think and to solve problems with particular emphasis on speed and efficiency of general understanding as well as on logical analytical thinking. Results: better than average.
c. Test of general ability to briefly concentrate while assessing the difference between meaningless signs quickly and accurately. Results: above average with a balanced relation between carefulness and speed.
d. Test for language-related learning and memory (including memory for texts, for meaningful relations and for numbers). Mr U's learning of lists of words was extremely quick and accurate. His repetition of spoken numbers forwards and back was above average, reflecting particularly on the executive functions of working memory. Also his auditory memory span in the working memory was above average for his age group, as was his very much attention-related short-term memory for numbers.

Thus, there are no indications of abnormal executive functions of his working memory.

e. Psychometric estimates of personality structure (Freiburg Personality Inventory) (FPI-A1) (standard results, averaging 7–3):

nervousness 7 (near above average);
spontaneous aggression 1 (extremely below average);
depression 4 (average);
excitability 1 (extremely below average);
sociability 6 (average);
equanimity 1 (average);
reactive aggression 3 (near below average);
repression 6 (average);
openness 3 (somewhat below average);
extraversion 4 (average);
emotional lability 3 (near below average); and
masculinity 2 (below average).

During the following question and answer period about the above results, his personality has been assessed to be prevailingly introverted. Furthermore, there were no indications of excitability or of limited frustration tolerance, and he was mostly in a balanced mood. No further singularities were detected.

f. Estimate of psychological disorders by the Minnesota Multiphasic Personality Inventory (MMPI – 2): *T* values (average span 30–70) Mr U's profile was as follows:

Scales of validity:
lying (*L*) 67 (above average);
rarity value (*F*) 52 (average);
correction value (*K* 63 (above average);

Clinical scales:
hypochondriasis 80 (above average);
depression 78 (above average);
hysteria 82 (above average);
psychopathy 74 (above average);
masculinity/femininity 52 (average);
paranoia 88 (above average);
psychasthenia 78 (above average);
schizophrenia 80 (above average);
hypomania 45 (average); and
social Introversion 44 (average).

Commentary: the *L* value (*T* = 67) reveals an initial resistance to the test.

The normal *F* value (*T* = 52) indicates that Mr U answered with the required honesty and openness. The above average *K* value (*T* = 63) generally indicates a denial of psychological weaknesses. The *K* scale requires a more detailed evaluation in combination with additional information about the tested person.

In general, the test reveals a difficult personality with abnormal affective reactions to experiences and with multiple somatic complaints. His schizophrenia scale includes unusual experiences and disturbed social contact. However, this scale cannot diagnose schizophrenia.

g. Estimate of important aspects of the personality, meaningful for the psychological personality structure (emotional mood and aspects of self experience, as well as psychosocial relations (using the Giessen test). This is a self-evaluation test revealing six personality traits. Counted are *T* values: *T* below 40 (below average), *T* 40–60 (average) and 61+ (above average):

social resonance (negative vs. positive social resonance) 47 (average);
dominance (in contrast to yielding) 44 (average);
control 45 (average);
basic mood (hypomanic vs. depressive) 50 (average);
forgetting (forgetting vs. retentive) 50 (average); and
social ability (socially potent vs. socially impotent) 41 (still average).

Commentary: there are no deviations from the norm. Mr U appears to have balanced control over social affect and impulses.

h. International Personality Disorder Examination (IPDE) (traits and behaviors important criteria for personality disturbances relevant to the classification systems ICD-10 and DSM-IV):

A screening questionnaire for self-evaluation is used containing 59 questions to be answered with 'true' or 'false'.
Based on the answers, there are merely certain hints of a very slight accentuation of single personality traits. Correspondingly, in the subsequent short interview there were merely insignificant indications of a mainly insecure, anxious personality with very few schizoid traits

i. Narcissism inventory:

Based on a questionnaire with 18 scales, Mr U showed a tendency toward above average narcissistic personality traits (*T* > 57), e.g., grandiose ideas regarding himself, while feelings of incompetence and of low self esteem were below average (*T* < 43).

j. The Toronto-Alexithymia Scale (regarding disturbances of experiencing emotions):

The psychometric scales suggest a generally difficult personality with several difficulties dealing with emotional experiences. Mr U's mood appears tense, anxious, suspicious and depressed. Initially he mentions many somatic complaints. He appears to be ruminating, introverted, now and then agitated. His emotional relations appear controlled and superficial with a tendency to blame others for his problems and to be unforgiving. However, there is no increased potential toward aggression. He considers himself as caring and as capable of lasting relationships.
Based on psychological testing, he shows specific disturbances in dealing with affect. He has difficulties in describing as well as in understanding emotions. Such limitations also include the physiological components of emotions.
Mr U presents, however, no disturbances in general problem-solving and should therefore be capable of utilizing experiences to evaluate intentions.

(A3) Reports on routine physical and neurological examinations (neurological history includes closed head injury with loss of consciousness, based on records and on patient's reports)

Reportedly, Mr U had generally been in good physical health. His routine physical and neurological examinations, as well as his laboratory tests, had all been within normal limits.

His birth (as an unwanted 10th child), his infancy, childhood and all developmental milestones had been 'completely normal', according to Mr U and his mother. He over-emphasized in a slightly defensive way that he had always been treated well by his parents, especially by his 'much too good-natured' father who had died of a heart attack 3 years ago (1 month before the birth of Mr U's son, one of his two victims). His mother is alive and well. One of his sisters had reported that their mother had avoided physical tenderness.

At age 20 he was involved in a severe frontal car accident with loss of consciousness for ca. 45 min, diagnosed as a comotio cerebri. A written witness report stated that Mr U's 'personality had changed' after this accident.

Mr U himself commented that after this car accident he had experienced about 16 *de novo* sensations of 'side to side motion with headaches'. Furthermore, there had been brief episodes 'like being in a plastic bag, not being aware of anything happening around, as if the plug had been pulled from an electric outlet'.

In addition, he related that a few years ago he had suffered 'a kind of a stroke' during which he had briefly 'lost the sense of touch all over the right side of the body' associated with certain difficulties expressing himself verbally. His own 'explanation' was that around that time his work had involved the handling of 'poisonous solvents'.

Commentary: some of his descriptions are consistent with a left-sided 'stroke' and/or might have been indicative of a focal seizure (Loddenkemper & Kotagal, 2005).

(A4) Objective brain tests

(A4a) Standard Scalp EEG. The examining physician had taken note of an 'undetermined syncope in 2000', based on the patient's description. At the time of these tests, the patient was without symptoms.

Findings: the unipolar leads separating the ears show a bilateral occipital well-presented basic 9/s alpha activity with satisfactory modulated amplitudes of ca. 40 μV. The remaining leads show a more reduced basic alpha rhythm, repeatedly including 7/s intermediate waves.

The bilateral parieto-occipital leads also show a moderate to well-developed basic alpha activity.

The blocking effect upon eye opening is positive. In between there are multiple muscular artifacts. Central EEG tracings show no physiological potentials attributed to technical conditions.

The transverse parieto-occipital leads show basic alpha activity, which is however decreased and with low amplitudes. The on–off effect continues to be positive.

The vertex lead shows again higher amplitudes between 50 and 60 μV. There is again an increased presentation of intermediate wave activity; however, not to the degree of dysrhythmia.

During hyperventilation (every 3 min) there is no measurable change in comparison to the rest-wake EEG. After completion of the hyperventilation there is a stretch with 5/s waves lasting 6/s. The amplitudes are higher over the right hemisphere than over the left one and measure 60 μV on the right side.

Under photo stimulation there is an increase of the flicker frequency synchronous to the phenomenon of 'photic driving', an abnormal rhythm of the discharge frequency. During continued photo stimulation numerous motion artifacts interfering with the interpretation.

Summary: during the rest-wake EEG there is a moderate to well-presented basic alpha rhythm activity. There is no change during hyperventilation. After hyperventilation there is a dysrhythmic EEG sequence. Under photo stimulation there are abnormal EEG rhythms There are, however, no indications of a focus. There is some question of differential hemispheric involvement, but no potentials indicative of epilepsy.

Interpretation: his scalp EEG has to be judged as 'borderline pathological'. The provocative procedures render the electrophysiological brain activity labile. In combination with the anamnesis this could be interpreted as a mild brain dysfunction under stress. During rest-wake conditions there are no pathological waves. The suspected brain dysfunction under provocation needs further clinical investigation.

(A4b) Cranial-CT scan. Question: Are there any morphological changes of the brain reflected on CT scan relevant to the determination of culpability?

The CT procedure involved native axial slices, 4-mm thick infratentorial and 8-mm thick supratentorial, represented in soft-tissue and in bony tissue images.

The right temporo-occipital portions of the calotte were less indented than those of the left side. Otherwise there were no bony deviations of the calotte. The depicted paranasal cavities as well as the mastoid cell systems showed normal pneumatisation.

The midline structures were not displaced. The attachment of the falx cerebri was significantly displaced to the right of the paramedian. This is associated with a hemispheric asymmetry with displacement of the right occipital regions. The left hemisphere showed no abnormality. There were no visible differences between the thickness of the cerebral cortex, basal ganglia and brain stem. There were no demonstrable focal defects in the regions of the cerebral cortex, basal ganglia and brain stem. The right ventricle in the region of the sella media as well as in that of the anterior horn presented more to the left than normal. There were no indications of an acute hindrance of the liquor circulation or flow. The basal cisterns are normally presented. The radiologist concluded that there was no direct or indirect indications of a space-occupying lesion.

Interpretation: hemispheric asymmetry, in particular, in regard to the reduced presentation of the right temporo-occipital region compared to the left side was noted. Because hemispheric asymmetries are frequently found, as a rule they are not considered to be pathological, the less so in the absence of focal defects. Beyond that, this is an age-appropriate unremarkable CT finding without indication of a space-occupying lesion.

(A5) Neuropsychiatric comments to EEG and CT-scan (on record)

In an overview of both objective brain tests, a neuropsychiatrist pointed to the borderline pathological signs within the context of the patient's history, including his head injury. He emphasized that such an injury frequently involves subtle damage to the temporo-limbic brain structures. The commentator continued:

> Notable was that both EEG and CT scans show abnormalities which included the occipital region. For example, the EEG revealed a reduced parieto-occipital basic activity, possibly related to the patient's head injury.

Also remarkable is the abnormal EEG rhythm ('photic driving') after photic stimulation and higher amplitudes between 50 and 60 µV in the vertex leads. Also the intermediary wave activity is higher than usual, albeit not to the extent of dysrhythmia. The commentator noted, however, that even such changes can be rarely detected, e.g., when they occur in deeper brain regions, not accessible by scalp EEG. Further, congruent with the EEG expert's findings of brain dysfunction upon provocation, the commentator recommended a T2 weighted MRI, particularly of the temporo-limbic brain structures. He added, however, that even an MRI cannot detect possible microscopic or neuro-electro-chemical changes, generally present during partial seizures. Therefore, 'neuro-psychiatric examinations are still considered to offer the most information' (as also emphasized by Geschwind, 1984, in regard to TLE). (Note, that even though an MRI had been recommended, forensic circumstances (detention in a maximum security facility) have prevented such additional examination.)

(B) New additional examinations (court-approved)

(B1) Neuropsychiatric Interviews (including exploration of subjective meaning of events, experiences and memories)

Mr U a right-handed, previously non-aggressive, separated white man, ca. 30 years of age, IQ 124 +, was fully employed in his field as an engineer (M.Sc.). He had been separated from his wife at his request 'for general incompatibility' and emphasized that even after the separation his wife had kept referring to him as 'a wonderful husband'. His verbal expression was fluent and coherent and without formal thought disturbances. No evidence of active hallucinations or delusions was detected. His prevalent facial expression conveyed the impression of a prematurely aged child, frequently presenting a somewhat forced joviality. He readily related to the interviewer (AAP), emphasizing that he felt comfortable talking to 'somebody with long life experience'. Mr U tended to express his emotions indirectly in the form of motor restlessness, suddenly jumping up and pacing the floor when talking about having killed his two adored children, acts incomprehensible to him. Utterly perplexed, he spoke about the contrast between his acts and his life-long care and worries about 'small children world wide with their helpless eyes'. Somewhat defensively he re-emphasized that his own child-hood had been 'completely normal' and that he 'had always been treated well'. However, since childhood he knew from an older sister that he had been an unwanted 10th child, conceived after his mother had had requested and received a 'steriliza-tion'. As his first childhood memory he remem-bered having been frightened by cold scissors suddenly placed on his head by his mother, without warning, 'to open a boil' (raising some questions about his mother's child caring style).

(B1a) Events preceding the double infanticides (elicited from the patient during the neuropsychiatric interview). On the fateful morning, Mr U and his two infants, who had been on a weekend visit with him, had 'all slept and felt well'. As he was caring efficiently for his infants, he suddenly noticed an earring on his 3-year-old son. He began to have strange epigrastric discomfort and sudden diarrhea.

Spontaneously he elaborated that the earring was a symbol that had revived 'a thousand memories of general (not extreme) neglect of children world wide'.

Comment: Mr U's subsequent bizarre changes in experience and behavior strongly suggest that the sight of the earring constituted his ultimate indi-vidualized trigger stimulus in a preceding intermit-tent sequence of stimuli (such as his memories of media reports) about neglected children. Such a constellation is a characteristic prerequisite for the neurophysiological mechanism of seizure kindling (Goddard, 1967; Goddard & McIntyre, 1986), an implication further supported by Mr U's subse-quent anomalous behavior.

(B2) Three seizure-like phases

(B2a) Aura with Gradual Progression into Ictus (After ca. 5 min). Mr U's suddenly unusual epi-gastric pain was followed by uncontrollable diarrhea. In the bathroom mirror he suddenly saw his face 'distorted, like a loser, like somebody who lost control over his life and over his ability to think'. This impression had been associated with paranoid thoughts: 'They have done me in'. 'They' referred to 'envious people everywhere who envied me my good job'. Further, 'they' also included his separated wife and the two homosexual men with whom she happened to share an apartment. These men had assumed much of his infants' care, and had not even invited him to the past christening of his 1-year-old daughter. Next, he had (and still) believed 'they have not only done me in but my

children as well'. As an example he thought of the two men's dangerous yet unrestrained pet dog which had recently bitten his son in the face. He began to recall news stories about child neglect in general and specifically in regard to his wife's abortion, about which he had been informed only recently. Further, to Mr U any kind of body piercing implied the danger of infection, to which his son had been particularly prone, including 'infections of his ear'. As briefly mentioned, around the time of the acts, all such memories and media reports related to potential neglect of children, together with similar 'thousand memories hurt like stabs in my head'. In vain he had shouted: 'Stop! Stop!' as his son's earring had increasingly become the sole focus of his attention.

(B2b) Escalation of aura into ictus. Mr U began to experience various additional visual illusions and hal-lucinations, especially distortions of the human face.

a. Initially (still during the aura?) he saw his own 'distorted face in the bathroom mirror'.
b. Then; 'I looked like through a box, everything right or left of it was extremely vague'.
c. 'My son's earring was so extremely bright, glis-tening like nothing real, unless from a laser, or a strange energy source. The brightness of the earring sort of pierced my eyes'.
d. 'Both my children's faces had no eyes or no facial features'.
e. 'I saw the head of my daughter (lying near her brother). Her head was about double its size, it looked cheesy-white, yellowish at places, bloated by disease, and had no facial features'.

Additional psychotic experiences, including deper-sonalization, occurred during the reported suffoca-tion of both his infants, for which he had used nearby plastic bags in plain sight (intended for computer parts).

f. 'I experienced myself as looking on without feeling anything, while someone else was doing it' (the killing).
g. 'Somehow I suddenly saw some things like spiders and I got the idea I have to save the children, I have to resuscitate the children', and he attempted to do so.

(B2c) Post-ictus. After his infanticides Mr U falsely 'confessed' by phone to his only friend and

later to the police, that he had killed his two children by 'breaking their necks when falling downstairs with them and turning their necks around'. (There is, however, no pathological evidence for such actions, only for suffocation.) After he had realized that he was unable to resuscitate his children, he had experienced 'something like panic and looked in vain for a telephone' to call a doctor. Then, 'somehow' he found himself with an electrical cord in his hands, about which he 'somehow knew' that it was too short to be put into the bath tub to kill himself. Thereafter he puzzled over the fact that he, having been trained in electro-engineering, had not even been able to use an extension cord that had been in plain view.

Comment: such strikingly inefficient behavior would be congruent with post-ictally lingering prefrontal dysfunction. Such a dysfunction would be consistent with the hypothesized brief fronto-limbic imbalance during LPTR, as would be an implicated lingering limbic (hippocampal?) dysfunctional activation, associated with distorted memory: The day before the incident, Mr U had actually almost fallen downstairs with both infants in his arms.

DIAGNOSTIC IMPRESSION

Atypical psychosis (DSM-III) or 'Psychotic Disorder due to a General Medical Condition (implicating partial limbic seizures) with Hallucinations and Delusions'. Appearing in Paroxysmal Episodes (DSM-IV # 293), vague diagnoses without heuristic value.

Although not used in previous reports on Mr U, a similar vague and not sufficiently differentiated category could be considered: 'Acute transient and psychotic disorders' ICD-10: F23 (Marneros & Pillman, 2004), proposed for ICD-11 (Jaeger, Riedel, & Moeller, 2007), although these symptoms last much longer than a few minutes.

More specifically, Mr U presents the proposed partial limbic seizure LPTR. The specific pattern of 13 interrelated symptoms and signs is present (Table 1), meeting all 16 inclusion and 13 exclusion criteria (Pontius, 1997; Pontius & Wieser, 2004, Table 1).

Alternative interpretations

Two previous psychiatric examiners and a psychologist's 10 neuro-psychological test results had merely diagnosed 'alexithymia' and 'dissociative personality' (without specifying organic vs. functional etiology). All such vague diagnostic impressions did not meet the legal requirements for 'insanity'. (Note, however, that neither one of these findings and diagnoses would be incompatible with LPTR.)

Optional: alternatively, it could be argued that Mr U is a psychopath, killing his children out of spite: if I cannot enjoy caring for them, nobody else should. In contrast to such a facile 'common sense' assumption, Mr U met all the specific criteria for LPTR, entirely opposite to psychopathy. Moreover, his visual illusions and hallucinations (macroscopia, metamorphosia and tunnel vision) implicate organicity (hardly to be invented by a malingering lay person). Most of all, psychopathy can be ruled out by Mr U's voluntary confessions and his assumption of full subjective responsibility and culpability with feelings of remorse, guilt and with suicidal ideation.

DISCUSSION

Dangerousness and potential likelihood to re-encounter the specific trigger stimuli

LPTR patients seem to present more danger to strangers (who unwittingly happen to present the individualized external trigger stimulus), than to continuously present family members (see 'Exceptional: Case' of grand-parental filicides, above). Such a preference for strangers might be explained by the difference between kindling and tolerance. Tolerance tends to develop toward continuously present stimuli, such as family members and is considered to be the reverse to kindling (Post & Kopanda, 1975). Kindling requires merely intermittent exposure to stimuli (Corcoran, & Solomon, 2005; Goddard, 1967; Goddard & McIntyre, 1986).

Specifically, the degree of dangerousness in LPTR depends essentially on the likely frequency with which patients' individualized and externally present trigger stimuli (whose ominous role is not known to them) might be re-encountered by chance (mostly after intervals of months or years).

In contrast to LPTR's hypothesized altered state of 'paleo-consciousness', there is no serious danger to others posed by patients with complete loss of consciousness in generalized seizures, nor with clouding of consciousness, as in complex partial seizures (e.g., in TLE, Delgado-Escueta et al., 1981).

Prevention and potential treatment

a. The present patient's alexithymia within the context of having been an unwanted 10th child is consistent with the finding that so far all LPTR patients had been social loners. Prevention might therefore begin during the loners' childhood (e.g., by providing Big Brothers or Big Sisters, since eventually also female LPTR patients might become identified).

b. As to a potential pharmaceutical treatment, so far only a few reports are available, tending to recommend anticonvulsants (e.g., carbamazepine, potentially helpful over 18 years for the longest observed LPTR patient, Pontius, 1996). The use of anticonvulsants against seizure kindling instead of anti-psychotics has also been supported by mammalian experiments (Ebert & Loescher, 1999).

c. Specific cognitive psychotherapy is preventively applicable, as in the present patient.

The aim of such insight therapy is to (1) raise patients' awareness of the ominous implications of their potentially recurrent (aural?) symptoms and (2) impress on them the extreme urgency to remove themselves promptly (within ca. 5 min) from the very place of such re-occurrence. Such a prompt change of place is indicated because the specific and external triggering stimuli (required for kindling, Table 1) are most likely embedded in the surroundings of the recurrent (aural?) symptoms.

Such a tactic had been detected spontaneously by one 'serial murderer' and thereafter successfully practiced by him on two occasions over several years, thereby sparing two additional lives (Pontius, 2002).

By contrast, if seizures are due to lesions in the brain (Table 4), there is no possible escape from them. Once internally evoked seizures (often with convulsions) have started, they take their relentless course. Prodromal symptoms may merely allow such patients to seek a spot to cushion their inevitable potential fall.

Future research on LPTR

(1a) The prevalence of social loners might implicate certain genetic factors in kindling (as in primates, Wada, 1998), recalling Kagan, Reznick, & Snidman's (1988) suggestion of biological factors in 'childhood shyness'.

(1b) The large number of subjects needed for genetic research might ultimately become available if LPTR could be identified in many more non-felonious 'sleeper' cases of LPTR with equally sudden, bizarre, unplanned, motiveless, non-volitional and out-of character social misbehavior.

(2) Might borderline or dormant brain pathology suddenly become clinically manifest during seizure activation (e.g., through kindling), as the present case history suggests?

TABLE 4
Database for case of limbic psychotic trigger reaction

A certain explanation might be indicated for the 24th case of the proposed paroxysmal symptomatology of LPTR, in regard to the typical database of 'Neurocase' typically with brain lesions. In contrast, the specific LPTR symptomatology (Table 1) implicates a partial non-convulsive behavioral limbic seizure, elicited by the neurophysiological mechanism of seizure kindling. This mechanism, however, requires no brain lesion, instead it is based on a primate model, where intermittent exposure to subthreshold stimuli kindles non-convulsive behavioral seizures. Thus,

(1) No mandatory brain lesions, congruent with experimental seizure kindling. The kindling effect, however, might be facilitated and/ or enhanced;
    (a) by a history of a closed head injury (as in the present case), and/or
    (b) by non-specific various, even borderline brain abnormalities on objective brain tests, (as by the present patient's borderline abnormalities in his temproro-parietal-occipital brain region on EEG and CT scan).
(2) Methods of assessment of paroxysmal LPTR are only available after the implicated seizure symptomatology, including the bizarre acts had passed;
    (a) three neuro-psychiatric interviews by three examiners exploring the subjective meaning of events and of memories,
    (b) 10 neuro-psychological tests,
    (c) assessment of social factors,
    (d) routine physical examination,
    (e) routine neurological examination, and
    (f) routine laboratory tests.
(3) Proposed diagnosis: LPTR in a 24th case, meeting 16 inclusion and 13 exclusion criteria. Its 13 symptoms and signs (Table 1) include out-of-character felonious acts (present patient's parental double infanticide). Such acts are motiveless, unplanned, unintentional, non-volitional, occurring without emotional involvement and are remembered afterwards.

If so, unexpected, even forensic consequences might ensue. Thus, Mr U's out-of-character acts (suffocation) might have been specifically facilitated by those visual illusions and hallucinations that distorted the faces of his beloved victims. Such kinds of visual perceptual distortions might have been specifically enhanced by his preexisting borderline posterior brain pathology (EEG/CT) with possible impact on the brain's specific face area (Puce, Allison, Gore, & McCarthy, 1995).

Alternatively, such objective findings might have been purely incidental, as possible within naturalistic settings.

Future research employing primate model

So far, experimental kindling has mostly employed rats, evoking convulsions (e.g., Corcoran & Solomon, 2005). As future research might employ more primates (monkeys) in whom kindling typically evokes non-convulsive behavioral seizures (Post & Kopanda, 1975; Wada, 1978, 1990), certain illuminating results might become available, extending the implicated analogy to LPTR patients.

With refinement of assessments and with enhanced technology, application of LPTR's primate model may offer a rather unique opportunity toward a certain objectification of non-voluntariness of action. Such a goal finds support by observation and patient report (Pontius, 2002), where such a drastic change from volitional to nonvolitional bizarre action could be rather pinpointed to occur only after the aura. To recall, during the aura the LPTR patient had still been able to act volitionally and effectively by escaping his seizure-related phenomena, thereby aborting a seizure recurrence (Pontius, 2002).

The ultimate goal of a certain objectification of loss of volition ('free will') may be attained during refined future monitoring. Thereby specific subtle changes in activity of certain brain regions may become detectable in behaving nonhuman primates (e.g., monkeys) during their kindled nonconvulsive behavioral seizures. It may be expected that a change-over phase from volition to nonvoluntariness occurs in analogy to LPTR between their aura (e.g., indicated by autonomic arousal and/or by signs of puzzlement) and their ictus (including bizarre behavior).

CONCLUSION

The present study of bizarre, otherwise unexplainable acts, attempts a certain explanation by implicating the neuro-physiological mechanism of seizure kindling, which can unite the unique 13 symptoms and signs (Table 1). These are determined by 16 inclusion and 13 exclusion criteria (Pontius & Wieser, 2004) and propose a partial limbic seizure: 'Limbic psychotic trigger reaction' as the proximate cause of the behaviors in question. The LPTR symptomatology is analogous to non-convulsive behavioral seizures with indications of visual hallucinations, kindled in primates (Post & Kopanda, 1975; Wada, 1978; 1990) (Table 1).

Certain future insight on the complex LPTR phenomenon might be gained by experimental kindling of primates, lately neglected in favor of convulsive kindling in rats (Corcoran & Solomon, 2005). Monitoring of experimentally kindled seizures in behaving primates by ongoing objective brain tests might be informative with respect to certain subtle degrees of differentiation between a primate's usual behavior and its gradual deviation from the norm occurring during non-convulsive behavioral seizures, as implicated in LPTR where loss of volitional action only occurred after the aura (e.g., Pontius, 2002). Thus, future elaboration of LPTR's primate model may present a quite unique opportunity to objectify certain aspects of volition ('free will').

REFERENCES

Adamec, R. E. (1987). Commentary to AA Pontius 'Psychotic Trigger Reaction': Neuro-psychiatric and neuro-biological (limbic?) aspects of homicide, reflecting on normal action. *Integrative Psychiatry, 5*, 130–134.

Aziz, M. A., Razik, G. N., & Donn, J. E. (2005). Dangerousness and management of delusional misidentification syndrome. *Psychopathology, 38*, 91–96.

Bear, D. M. (1987). Commentary to AA Pontius: Psychotic trigger reaction: Neuro-psychiatric and neuro-biological (limbic?) aspects of homicide, reflecting on normal action. *Integrative Psychiatry, 5*, 125–127.

Cain, D. P. (1992). Kindling and the amygdala. In J. P. Aggleton (Ed.), *The amygdala* (pp. 539–560). New York: Wiley.

Clore, G. (1992). Cognitive phenomenology: Feeling and the construction of judgement. In L. L. Martin (Ed.), *The construction of social judgements* (pp. 133–163). Hillsdale, NJ: Erlbaum.

Corcoran, M. E., Solomon, L. M. (Eds) (2005). *Kindling 6 (Advances in behavioral biology)*. New York: Springer.

Cromie, W. J. (1996). Sudden brain seizures said to trigger violence. *Harvard University Gazette, 41*, 5.

Cromie, W. J. (2002). Brain changes in learning measured. *Harvard University Gazette, 47*, 1–4.

Damasio, A. R. (1999). *The feeling of what happens. Body and emotion in the making of consciousness.* New York: Harcourt Brace.

Delgado-Escueta, A. V., Mattson, R. H., King, L., Goldensohn, E. S., Spiegel, H., Madson, J., Crandall, P., Dreyfus, F., & Porter, R. J. (1981). Special report: The nature of aggression during epileptic seizures. *New England Journal of Medicine, 305*, 711–716.

Devinsky, O., Kelley, K., Porter, R. J., & Theodore, W. H. (1988). Clinical and electroencephalographic features of simple partial seizures. *Neurology, 38*, 1347–1352.

Ebert, U., & Loescher, W. (1999). Pathophysiologie des Kindling-Phenomens. Ansaetze zur Entwicklung neuer Antiepileptika. [Pathophysiology of the kindling phenomenon: Towards the development of new antiepileptica]. *Neuroforum, 3*(99), 76–86.

Fried, I. (1997). 'Syndrome E'. *Lancet, 350*, 1845–1848.

Fuchs, T. (2005). Oekologie des Gehirns. [Oecology of the brain]. *Nervenarzt, 76*, 1–10.

Garland, E. (2004). *Neuroscience and the law.* Washington, DC: Dana Press.

Geschwind, N. (1984). The clinician scientists. *Science, 224*, 243.

Geschwind, N., & Sherwin, I. (1967). Language-induced epilepsy. *Archives Neurology, 16*, 25–31.

Gloor, P., Olivier, A., Quesney, L. F., Andermann, F., & Horowitz, S. (1982). The role of the limbic system in experiential phenomena of temporal lobe epilepsy. *Archives of General Psychiatry, 12*, 129–144.

Goddard, G. V. (1967). Development of epileptic seizures through brain stimulation of low intentsity. *Nature, 214*, 1020–1021.

Goddard, G. V., & McIntyre, D. C. (1986). Some properties of lasting epileptogenic trace kindled by repeated electrical stimulation of the amygdala in mammals. In B. K. Doane & K. E. Livingston (Eds), *The limbic system: Functional and clinical disorders* (pp. 95–105). New York: Raven.

Halleck, S. L. (1992). Assessment of voluntariness of behavior. *Bulletin of the American Academy of Psychiatry & Law, 20*, 221–236.

Hanson, N. R. (1965). *Patterns of discovery.* Cambridge, UK: Cambridge University Press.

Harre, R. (1970). *The principles of scientific thinking.* Chicago, IL: University Chicago Press.

Hayek, F. A. (1964). The theory of complex phenomena. In M. Bunge (Ed.), *The critical approach to science and philosophy* (pp. 332–349). London: Free Press of Glencoe, Collier & Macmillan.

Heath, R. G., Monroe, R. R., & Mickle, W. (1955). Stimulation of the amygdaloid nucleus in a schizophrenic patient. *American Journal of Psychiatry, 111*, 862–863.

Jaeger, M., Riedel, M., & Moeller, H. J. (2007). Akute voruebergehende psychotische Stoerungen [Acute and transient psychotic disorders] (ICD-10: F23). *Nervenarzt, 78*, 745–752.

Kagan, J., Reznick, J. S., & Snidman, N. (1988). Biological bases of childhood shyness. *Science, 240*, 167–171.

Kandel, E. R. (2001). The molecular biology of memory storage: A dialogue between genes and synapses. *Science, 294*, 1030–1038.

Krishnamoorthy, E. S., & Trimble, M. R. (1998). Mechanism of forced normalization. In M. R. Trimble & B. Schmitz (Eds), *Forced normalization and alternative psychoses of epilepsy* (pp. 193–209). Petersfield, UK: Wrightson Biomedical Publishing Ltd.

Kuhn, T. S. (1977). *The essential tension: Selected studies in scientific tradition and change.* Chicago, IL: University of Chicago Press.

Kuzma, J. M., & Black, D. W. (2004). Compulsive disorders. *Current Psychiatry Reports, 6*, 58–65.

Landau, M. (1996). Explaining the unexplainable. *Focus (Harvard Medical School), 4–26*, 10–11.

Levin, H. X., Eisenberg, H. M., & Benton, A. L. (1989). *Mild head injury.* New York: Oxford University Press.

L'hermitte, F. (1986). Human autonomy and the frontal lobes. II. Patient behavior in complex and social situations. The 'environmental dependency syndrome'. *Annals of Neurology, 19*, 334–343.

Loddenkemper, T., & Kotagal, P. (2005). Lateralization signs during seizures in focal epilepsy. *Epilepsy & Behavior, 7*, 1–17.

LoPiccolo, P. (1996). Something snapped. *Technology Review (MIT), 99*, 52–67.

MacLean, P. (1990). *The triune brain in evolution: Role in paleocerebral functions.* New York: Plenum.

MacLean, P. (1992). The limbic system concept. In M. R.Trimble & T. G. Bolwig (Eds), *The temporal lobes and the limbic system* (pp. 1–13). Petersfield, UK: Wrightson Biomedical Publishing.

Marneros, A., & Pillman, F. (2004). *Acute and transient psychoses.* Cambridge, UK: Cambridge University Press.

Morrell, F. (1985). Secondary epileptogenesis in man. *Archives of Neurology, 42*, 318–335.

Nauta, W. J. H. (1971). The problem of the frontal lobes – A reinterpretation. *Journal of Psychiatric Research, 8*, 167–187.

Pontius, A. A. (1981). Stimuli triggering violence in psychoses. *Journal of Forensic Sciences, 26*, 123–128.

Pontius, A. A. (1984). Specific stimulus-evoked violent action in psychotic trigger reaction: A seizure-like imbalance between frontal lobe and limbic systems? *Perceptual & Motor Skills Monograph I-59* (Suppl.), 299–333.

Pontius, A. A. (1986). *Biology of violence, psychotic trigger reaction, co-organizer and talk, respectively.* Annual Meeting, Association for the Advancement of Science.

Pontius, A. A. (1987). Psychotic trigger reaction: Neuropsychiatric and neuro-biological (limbic?) aspects of homicide, reflecting on normal action. *Integrative Psychiatry, 5*, 116–139.

Pontius, A. A. (1990). Infanticide in limbic psychotic trigger reaction in man with Jacksonian seizures and petit mal seizures: Kindling by traumatic experiences. *Psychological Reports, 67*, 935–945.

Pontius, A. A. (1993a). Neuroethological aspects of certain limbic seizure-like dysfunctions: Exemplified by limbic psychotic trigger reaction (motiveless homicide with intact memory). *Integrative Psychiatry, 9,* 151–167.

Pontius, A. A. (1993b). Overwhelming remembrance of things past: Proust portrays limbic kindling by external stimulus. Literary genius can presage neurobiological patterns of puzzling behavior. *Psychological Reports, 73,* 615–621.

Pontius, A. A. (1993c). Neuropsychiatric update of the crime 'profile' and 'signature' in single or serial homicides: Rule out Limbic psychotic trigger reaction. *Psychological Reports, 73,* 875–892.

Pontius, A. A. (1995). Retroductive reasoning in a proposed new subtype of partial seizures evoked by 'limbic kindling'. *Psychological Reports, 76,* 55–62.

Pontius, A. A. (1996a). *The limbic system, limbic psychotic trigger reaction Organizer and talk respectively.* Annual Meeting, Association for the Advancement of Science.

Pontius, A. A. (1996b). Forensic significance of the limbic psychotic trigger reaction. *Bulletin of the American Academy of Psychiatry & Law, 24,* 125–134.

Pontius, A. A. (1997). Homicide linked to moderate repetitive stresses kindling limbic seizures in 14 cases of limbic psychotic trigger reaction. *Aggression & Violent Behavior, 2,* 125–141.

Pontius, A. A. (1999). Motiveless fire setting, implicating partial limbic seizure kindling by revived memories of fires in 'limbic psychotic trigger reaction'. *Perceptual Motor Skills, 88,* 970–982.

Pontius, A. A. (2000). Comparison between two opposite homicidal syndromes (syndrome E vs. limbic psychotic trigger reaction). *Aggression & Violent Behavior, 5,* 423–427.

Pontius, A. A. (2001a). Two bank robbers with 'antisocial' and 'schizoid-avoidant' personality disorder, comorbid with partial seizures: Temporal lobe epilepsy and limbic psychotic trigger reaction, *respectively. Journal of Developmental and Physical Disability, 13,* 191–197.

Pontius, A. A. (2001b). Homicides with partial limbic seizures: Is chemical kindling the culprit? *International Journal of Offender Therapy & Comparative Criminology, 45,* 515–527.

Pontius, A. A. (2002a). Serial murderer learns to regain volition by recognizing the aura of his partial seizures of 'limbic psychotic trigger reaction'. *Clinical Case Studies, 1,* 324–341.

Pontius, A. A. (2002b). Neuroethology, exemplified by limbic seizures with motiveless homicide in 'Limbic Psychotic Trigger Reaction'. In G. A. Cory Jr & R. Gardner Jr (Eds), *The evolutionary neuroethology of Paul MacLean* (pp. 167–191). Westport, CT: Praeger.

Pontius, A. A. (2003). From volitional action to automatized homicide: Changing levels of self and consciousness during partial limbic seizures. *Aggression & Violent Behavior, 8,* 547– 561.

Pontius, A. A. (2004). Violence in schizophrenia vs. limbic psychotic trigger reaction: Prefrontal aspects of volitional action. *Aggression & Violent Behavior, 9,* 503–521.

Pontius, A. A. (2005). *Bizarre motiveless (Grand-)parental fillicide, implicating partial kindled Seizure: 'Limbic psychotic trigger reaction'.* Talk, 20th Munich University Fall Meeting of Forensic Psychiatry.

Pontius, A. A., & LeMay, M. J. (2003). Aggression in temporal lobe epilepsy and 'limbic psychotic trigger reaction' implicating vagus stimulation of hippocampus/amygdala (in sinus problems on MRIs). *Aggression & Violent Behavior, 8,* 245–257.

Pontius, A. A., & Wieser, H. G. (2004). Can memories kindle nonconvulsive behavioral seizures in humans? Case report exemplifying the 'limbic psychotic trigger reaction'. *Epilepsy & Behavior, 5,* 775–783.

Post, R. M., & Kopanda, R. T. (1975). Cocaine, kindling and reverse tolerance. *Lancet, i,* 409–410.

Proust, M. (1913). *A la recherché du temps perdu.* Paris, France: Editions Gallimard 1954 (original Published 1913).

Puce, A., Allison, T., Gore, J. C., & McCarthy, G. (1995). Face sensitive region in human extrastriate cortex studied by functional MRI. *Journal of Neurophysiology, 74,* 1192–1199.

Racine, R. J. (1980). Kindling: The first decade. *Neurosurgery, 3,* 324–352.

Racine, R. J., & Burnham, W. M. (1982). The kindling model. In P. A. Schwartzkroin & H. Wheal (Eds), *Electrophysiology of epilepsy* (pp. 153–171). London: Academic Press.

Shostakovitch, B. V., & Leonova, O. V. (2004). The concept of the limbic psychotic trigger reaction and its forensic significance (analytical review). *Russian Journal of Psychiatry, 1,* 60–67.

Sramka, M., Sedlak, F., & Nadvornik, P. (1984). Observation of kindling phenomenon in treatment of pain by stimulation in thalamus. In W. H. Sweet, S. Abrador, & J. Martin-Rodriguez (Eds), *Neurosurgical treatment in psychiatry* (pp. 651–654). New York: Elsevier.

Wada, J. A. (1978). The clinical relevance of kindling: Species, brain sites and seizure susceptibility. In K. E. Livingston & O. Hornykiewicz (Eds), *Limbic mechanism, the continuing evolution of the limbic system concept* (pp. 369–388). New York: Plenum.

Wada, J. A. (1990). Erosion of kindled epileptogenesis and kindling-induced long-term seizure suppressive effects in primates. In J. A. Wada (Ed.), *Kindling 4* (pp. 383–395). New York: Plenum.

Wada, J. A. (1998). Genetic disposition and kindling susceptibility in primates. In M. E. Corcoran & S. Moshe (Ed.), *Kindling 5* (pp. 1–11). New York: Plenum.

Wieser, H. G. (1983). The phenomenology of limbic seizures. In H. G. Wieser, E. J. Speckmann, & J. Engel Jr (Eds), *The epileptic focus* (pp. 113–136). London: John Libbey.

Wieser, H. G. (1993). *Electroclinial features of the psychomotor seizures: A stereoelectrographic study of ictal symptoms and chronopographical seizure patterns including clinical effects of intracerebral stimulation.* Stuttgart/London: Fischer-Butterworths.

NEUROCASE
2008, 14 (1), 44–53

Cross-examining dissociative identity disorder: Neuroimaging and etiology on trial

A. A. T. Simone Reinders[1,2]

[1]King's College London, Institute of Psychiatry, Division of Psychological Medicine, London, UK
[2]BCN Neuroimaging Center, University Medical Center Groningen, University of Groningen, Groningen, The Netherlands

Dissociative identity disorder (DID) is probably the most disputed of psychiatric diagnoses and of psychological forensic evaluations in the legal arena. The *iatrogenic* proponents assert that DID phenomena originate from psychotherapeutic treatment while *traumagenic* proponents state that DID develops after severe and chronic childhood trauma. In addition, DID that is simulated with malingering intentions, but not stimulated by psychotherapeutic treatment, may be called *pseudogenic*. With DID gaining more interest among the general public it can be expected that the number of *pseudogenic* cases will grow and the need to distinguish between *traumagenic*, *iatrogenic* or *pseudogenic* DID will increase accordingly. This paper discusses whether brain imaging studies can inform the judiciary and/or distinguish the etiology of DID.

Keywords: Dissociative identity disorder; Multiple personality disorder; Neuroimaging; Iatrogenic; Traumagenic; Pseudogenic; Etiology; Aetiology; Court; Forensic.

GENERAL INTRODUCTION

Dissociative identity disorder (DID), better known as multiple personality disorder (MPD) among non-professionals, has been of great interest to the public for over more than a century. The classic example from early fiction is of course *Dr. Jekyll and Mr. Hyde*. The continuing strong fascination of the public[1] with multiple personalities can be found in the recent film version of the book *The Lord of the Rings*, where in the movie the duality of the character Smeagol/Gollum receives more emphasis than in the book. The most recent of examples is the obsession of the public with the *Harry Potter* series, in which several of the characters incorporate split personalities, e.g., Voldemort, who splits his soul into seven pieces; Harry, who houses one of them next to his own main personality; the two faces of Severus Snape and Quirrell; and Lupin in the form of a time-tested standard werewolf. Even celebrities in the music industry like Robbie Williams and Beyoncé speak about their alter egos who take care of public performances and the interaction with the media in protection of their personal identity.

The movie *Primal Fear* (1996) shows an example of how a suspect successfully malingers by simulating DID and escapes rightful punishment. Although this specific example is fiction, it reflects the problem in real life of whether the etiology of a case of DID is traumagenic, iatrogenic or pseudogenic in court and law (Behnke, 1997). In the last 15

---

[1]For an extended list of examples of DID in fiction: *http://en.wikipedia.org/wiki/DID/MPD_in_fiction*

A. A. T. S. Reinders is supported by the Netherlands Organization for Scientific Research (www.nwo.nl), NWO-VENI grant no. 451-07-009.
Address correspondence to A. A. T. Simone Reinders, MSc, PhD, King's College London, Institute of Psychiatry (IoP), Division of Psychological Medicine, De Crespigny Park, London SE5 8AF, UK (E-mail: a.a.t.s.reinders@iop.kcl.ac.uk or a.a.t.s.reinders@med.umcg.nl).

 DOI: 10.1080/13554790801992768

years especially, the number of forensic cases involving MPD/DID has increased substantially (Frankel & Dalenberg, 2006). In parallel, the number of published empirical studies has also increased, including those involving neuroimaging. In addition, DID is probably the most disputed of psychiatric diagnoses and of psychological forensic evaluations in the legal arena. How can the empirical studies concerning brain imaging contribute to the psychiatric field when (the etiology of) DID is still haunted with skeptics and what advice can be given to the judiciary from this neuroscientific point of view? This paper (i) provides definitions for traumagenic, iatrogenic (subconscious or conscious) and pseudogenic DID, (ii) discusses if forensic psychiatry can benefit from these empirical studies in categorizing whether a case of DID is traumagenic, iatrogenic or pseudogenic, and (iii) identifies which concerns should be kept in mind when considering neuroimaging during forensic psychiatric/psychological evaluations.

THE IATROGENIC POSITION

The iatrogenic proponents assert that DID is manufactured during (possibly suggestive) psychotherapeutic treatment, often going hand in hand with the creation of false memories and the creation of separate and distinct identities. The iatrogenic position does include a reference to memories of a traumatic past, but it states that these are manufactured during psychotherapy, while in contrast the traumagenic view is based on the traumatic past being genuine. Indeed, from an iatrogenic view the very existence of dissociative identity states is questioned. This often leads to a biased presentation of facts, research, and/or literature (Gleaves, 1996; Pope, Barry, Bodkin, & Hudson, 2006; Spanos, 1994). Representative of the iatrogenic position, including a literature review, are papers by Piper and Merskey (2004a, 2004b, 2005), which were followed by consecutive responses from proponents of the traumagenic view (Coons, 2005; Fraser, 2005; Sar, 2005). This shows that even after the official recognition in the DSM-IV (DSM, edition IV, American Psychiatric Association, 1994), a consensus on the diagnosis and therapy of DID remains elusive and supporters of the diametrically opposed iatrogenic and traumagenic position still engage in passionate debate.

Actually, within the iatrogenic view two situations can be distinguished: (i) subjects emulate DID on a subconscious level, or (ii) subjects emulate DID on a conscious level. In the first case, subjects are convinced (i.e., by the therapist) that they suffer from DID. Consequently, they believe that they consist of multiple personalities without having conscious control over their situation. Therapeutic dependency, high suggestibility or high fantasy proneness may contribute to the effectiveness of the therapeutic intervention in the creation of DID phenomena. In this case, subjects genuinely believe that they suffer from DID. Therefore, this is a subconscious process as they are not aware of their simulation, and in that sense the therapist bears full responsibility for the generated DID. In the second case, subjects actively simulate DID through conscious intervention to satisfy their therapist. Therefore, displaying DID symptoms is consciously chosen behaviour, but without any ulterior motive other than to please the therapist. In a more general context, conscious DID simulation is referred to as malingering and is therefore described in the pseudogenic section below. Conscious DID simulation is also increasingly included as a control situation in empirical DID research (e.g., Huntjens, 2003).

One of the best known DID cases, *Sybil* (Schreiber, 1973), has been found (Rieber, 1999) to have been a manufactured iatrogenic case of multiple personality. Rieber (1999) discovered how easily 'The fine line between self-deception and deception of others' can be crossed when wanting to make a dissociative identity disorder case 'no matter what'. *Sybil* was manufactured through hypnosis, pentothal and a close emotional involvement between subject and therapist. Using the subject's given capacity for suggestion, false memories of sexual abuse were created and the notion of multiple personality implanted.

THE TRAUMAGENIC POSITION

When a subject is diagnosed with DID according to the official Diagnostic and Statistical Manual (DSM), edition IV (American Psychiatric Association, 1994) the case is referred to as genuine DID. According to the DSM-IV, DID is characterized by, among others, a disruption in consciousness and the presence of two or more distinct identities or personality states, also referred to as 'different emotional states', 'alters' or 'dissociative identity states'. Each of these states has its own perception of, relating to, and thinking about the environment and the self (Reinders et al., 2003).

Although these official criteria for DID do not specifically indicate a traumagenic origin (Gleaves, May, & Cardena, 2001), a history of childhood trauma is assumed to have caused the disorder. More specifically, traumagenic proponents hold that DID constitutes a severe form of post-traumatic stress disorder (PTSD). Within the traumagenic position, DID is believed to originate from a sub-conscious self-protecting reaction in order to cope with chronic, severe trauma such as persistent abuse (Nijenhuis, van der Hart, & Steele, 2002; Stickley & Nickeas, 2006). According to this traumagenic model, a subject can dissociate (i.e., mentally compartmentalize; Brown, 2006) from their painful experiences by repressing (i.e., inducing amnesia) the memories of these experiences. Dissociation and repression keep trauma-related memories out of conscious awareness to avoid (i.e., protect from) intolerable psychological distress. In this manner different personality states are created.

This theory is supported by numerous correlational studies (Bowman, Blix, & Coons, 1985; Dalenberg & Palesh, 2004; McElroy, 1992; Nijenhuis, Spinhoven, van Dyck, van der Hart, & Vanderlinden, 1998; Ross, Norton, & Wozney, 1989; Zlotnick et al., 1996). These studies usually rely on self-reports and confirm a high incidence of childhood trauma, e.g., childhood sexual abuse, physical neglect and/or emotional abuse, in adults and children with high levels of dissociation or DID. These studies are supported by recent epidemiological studies (Sar et al., 2007a; Xiao et al., 2006), which investigate the prevalence of DID and the relation to traumatic experiences in the general population. In support of the traumagenic position these studies found that pathological dissociation is more prevalent in more traumatized subsamples.

Following previously used descriptions and terminology (Reinders et al., 2003, 2006, i.e., avoiding interpretive discussions, cf. Merckelbach, Devilly, & Rassin, 2002), the dissociative identity states are indicated here as neutral identity states (NIS) and trauma-related identity states (TIS). In a NIS, DID subjects concentrate on functioning in daily life. To that end, NIS has amnesia for traumatic memories, thereby disabling recognition of the self-referential nature of trauma-related information. In contrast, the TIS does have conscious access to the traumatic memories (Dorahy, 2001; Van der Kolk & Fisler, 1995). These states do not know of each other's existence unless their treatment has progressed to phase II (Brown, Scheflin, & Hammond, 1998;

Steele, van der Hart, & Nijenhuis, 2001).[2] The sub-conscious co-existence of, and switching between, different states leads to reports of black-outs and amnestic periods.

Although holders of the traumagenic view usually do not deny that some symptoms or phenomena of DID can be created iatrogenically, they do stress that there is no evidence to suggest that the disorder itself can truly be created (Gleaves, 1996). On the other hand, even if DID symptoms can be created iatrogenically or enacted (Spanos, 1994) it does not mean that genuine traumagenic DID does not exist (Elzinga, Van Dyck, & Spinhoven, 1998).

THE PSEUDOGENIC POSITION

Pseudogenic[3] DID includes subjects who are simulating DID without any therapeutic intervention. This is a conscious and active simulation process for secondary gain. Several and diverse reasons can underlie this active, malingering DID simulation. Needy and attention seeking behaviour (histrionic) is generally relatively harmless. However, other reasons can be less innocuous: obtaining financial, social welfare or legal benefits (Rogers, 1997). For example, by feigning DID an accused can try to deceive a jury and judge in the hope to be held not accountable or responsible for the crimes committed thereby avoiding legal consequences, especially incarceration or execution.

DID AND JURISPRUDENCE

Merckelbach et al. (2002) distinguishes three categories of legal complications. The *first* category consists of DID subjects who start to accuse another person of severe and chronic (sexual) abuse during the subject's childhood. Almost always, objective evidence (e.g., diaries, eye-witness testimonies, written confessions or photographs taken by perpetrator) of the abuses is lacking (Piper & Merskey, 2004a). Without objective evidence the only remaining 'proof' is the (recovered) memories of childhood abuse.

---

[2]Phase II is an advanced phase of treatment which involves therapeutic exposure to traumatic memories and allows for self-initiated and self-controlled switching between dissociative identity states.
[3]Pseudo: fake; falsely; sham; feigned; deceptive resemblance to a specified thing.

A recent review by Laney and Loftus (2005) describes how memory is malleable and false memories can be planted, e.g., by the therapist. False memories (see for review and ethical issues: DePrince, Allard, & Oh, 2004) often include recovered memories, which are memories for which a subject claims to have been amnestic. The period of time being amnestic, i.e., the impressions of previous non-recall, has been shown to be often highly unreliable (e.g., Merckelbach et al., 2006), revealing a severe problem with the interpretation of amnesia. Interestingly, recent evidence shows that subjects who report recovered childhood trauma-related memories are more prone to falsely recalling and recognizing words (Geraerts, Smeets, Jelicic, van Heerden, & Merckelbach, 2005; Geraerts, 2006). Intriguingly, this was significantly correlated with fantasy proneness (as measured by the Creative Experiences Questionnaire, CEQ; Merckelbach, Horselenberg, & Muris, 2001) and not with dissociative symptoms (as measured by the Dissociative Experience Scale, DES; Bernstein & Putnam, 1986). Although this research did not include subjects diagnosed with DID, it supports the claims of the iatrogenic view that the creation or recovery of false memories is dependent on the level of hypnotizability, suggestibility or fantasy proneness (Merckelbach & Muris, 2001). Caution should therefore be taken when assuming a simple and direct causal relation between self-reported trauma and dissociation (Merckelbach & Muris, 2001).

The *second* category, from which legal complications may arise, is when a DID subject claims not to be responsible for crimes committed because a different identity state committed the crime and the dominant identity state claims not to be consciously aware of the crime due to inter-identity amnesia. This might hold in non-pseudogenic cases as long as the dominant identity state lacks both awareness of, and control over, the actions of that other identity state (Deeley, 2003). The *third* category (related to the second) has to do with civil rights. If 'alters' are interpreted as genuine entities, which alter serves as the legal representative of that person? Thus, who of the alters can sign a contract, give informed consent in the case of participation in a research project, or vote?

NEUROIMAGING THE ETIOLOGY OF DID

Imaging traumagenesis or iatrogenesis

Increasing amounts of empirical data directly or indirectly related to study the etiology of DID is appearing in the literature. By far the most seductive development is the application of neuroimaging in psychobiological DID research. Structural magnetic resonance imaging (sMRI; Tsai, Condie, Wu, & Chang, 1999; Vermetten, Schmahl, Lindner, Loewenstein, & Bremner, 2006), functional magnetic resonance imaging (fMRI; Tsai et al., 1999), positron emission tomography (PET; Mathew, Jack, & West, 1985; Reinders, 2004; Reinders et al., 2003, 2006), single photon emission computed tomography (SPECT; Sar, Unal, Kiziltan, Kundaci, & Ozturk, 2001; Sar, Unal, & Ozturk, 2007b; Saxe, Vasile, Hill, Bloomingdale, & Van der Kolk, 1992; Sheehan, Thurber, & Sewall, 2006), event-related potentials (ERPs; Allen & Movius, 2000) and electroencephalography (EEG; e.g., Coons, Milstein, & Marley, 1982; Hughes, Kuhlman, Fichtner, & Gruenfeld, 1990; Lapointe, Crayton, DeVito, Fichtner, & Konopka, 2006; Mesulam, 1981) have been applied to chart the psychobiological constituents of DID. Details on scanning techniques, the numbers of participants (healthy controls and DID subjects), physical gender, the diagnostic tools used to diagnose DID subjects, the DIS in which the DID subject remained during data acquisition and the experimental task(s) are listed in Table 1. This table clearly shows that no convergence exists the study of DID (Barlow, 2007). Table 2 lists findings from the neuroimaging studies listed in Table 1 (independent of task, identity state, activation or deactivation, etc.) in terms of brain lobes and sites, and shows that brain research in DID has produced a wide diversity of results. Combined with the lack of convergence on, e.g., brain mapping technique, experimental task or identity state under investigation, it is difficult to speculate about the brain mechanisms involved in DID. In addition, most of these studies are performed from the traumagenic position and lack the explicit investigation of possible iatrogenic influences, i.e., none of these studies have addressed the possibility of subconscious iatrogenically induced DID.

Four studies have surpassed the limitations of case reports and accepted the challenge of including a larger group of DID subjects. The study of Sar et al. (2007b) included 21 genuine DID subjects and nine healthy controls, the study of Sar et al. (2001) included 15 genuine DID subjects and eight healthy controls, the study of Vermetten et al. (2006) included 15 genuine DID subjects and 23 controls and the study of Reinders et al. (2003, 2006) included 11 genuine DID subjects. Besides the strength of including a larger group of subjects,

TABLE 1
Parameter review of neuroimaging studies on DID

Technique	References	N controls (M/F)	N DID (M/F)	Diagnostic tool	N DIS (type)[a]	Task(s)
sMRI	Vermetten et al. (2006)	23 (F)*	15 (F)	SCID-D DSM-IV	Unknown	n.a.
	Tsai et al. (1999)	52 (27/25)*[b]	1 (F)	DSM-IV	Unknown	n.a.
fMRI	Tsai et al. (1999)	0	1 (F)	DSM-IV	2 (D+A)	DIS switching
PET	Reinders et al. (2003, 2006)	0	11 (F)	SCID-D DSM-IV	2 (NIS+TIS)[c]	Memory[d]
	Mathew et al. (1985)	3 (F)*	1 (F)	DSM-III	3 (1 D+1 C & 1 I)	Rest
SPECT	Sar et al. (2007b)	9 (3/6)*	21 (7/14)	SCID-D DSM-IV+DES	1 (D)	Rest
	Sheehan et al. (2006)	0	1 (F)	None	2 (1 D+1 C)	Unknown
	Sar et al. (2001)	8 (2/6)*	15 (4/11)	SCID-D DSM-IV+DES	2 (1 D+1 A) or 1 (D)[e]	Rest
	Saxe et al. (1992)	0	1 (F)	Unknown	4 (unknown)	Rest
ERP	Allen et al. (2000)	60* (unknown)	4 (F)	SCID-D DSM-IV	2 (unknown[c])	Memory
EEG[f]	Lapointe et al. (2006)	4* (F)	3 (F)	DSM-III-R	2[g]	Rest
	Hughes et al. (1990)	1 (F)**	1 (F)	DSM-III-R	11 (1 D+10 A)[h]	2 tasks[i]
	Coons et al. (1982)	1 (M)**	2 (F)	Unknown	4/5 (1 D+3/4 A)[j]	7 tasks[k]
	Mesulam et al. (1981)	0	7 (F)	None	Unknown (unknown)	4 tasks[l]

sMRI, structural magnetic resonance imaging; fMRI, functional magnetic resonance imaging; PET, positron emission tomography; SPECT, single photon emission computed tomography; ERP, event-related potential; EEG, electroencephalogram; M, Male; F, Female; N, number; DID, dissociative identity disorder; DIS, dissociative identity state; NIS, neutral identity state, not aware of traumatic past; TIS, traumatic identity state, aware of traumatic past; SCID-D DSM-IV, structured clinical interview for DSM-IV dissociative disorders (Steinberg, 1993); DES, dissociative experience scale (Bernstein et al., 1986).

*Healthy control subject(s) did not mimic DID subject's identity state(s).

**Healthy control subject(s) did mimic DID subject's identity state(s).

[a]Number and type of DIS present during the measurement(s). The dominant (D) DIS refers to the personality state that covers the longest part of time throughout an ordinary day of the DID subject. The D DID is also referred to as native (Tsai et al., 1999), designated (Mathew et al., 1985), host (Sar et al., 2001, 2007b and Lapointe et al., 2006), adult (Sheehan et al., 2006), basic (Hughes et al.,1990) or primary (Coons et al., 1982, Mesulam et al., 1981). An alter (A) is a sub-dominant DIS and a child (C) alter has a subjective age of a young child. The integrated (I) state is the identity after the treatment.

[b]From Stern et al. (1996), i.e., literature comparison only.

[c]Report inter-identity amnesia for test material.

[d]Both NIS and TIS listened to an autobiographical neutral and trauma-related text in a symptom provocation paradigm.

[e]Six DID subjects were scanned once in D and twice in the same A, 9 DID subjects (D only) and the healthy controls were scanned once.

[f]Most commonly cited references.

[g]DID subject 1: DIS 1 = F (C); DIS 2 = F
DID subject 2: DIS 1 = unknown; DIS 2 = unknown
DID subject 3: DIS 1 = M; DIS 2 = F (D)

[h]2 C, two adolescents, seven adults (1 D).

[i]Eyes open and eyes closed.

[j]DID subject 1 and the control: 4 DIS (3 F (1 D, 1 C), 1 M)
DID subject 2: 5 DIS (3 F (1 D), 2 M)

[k]Awake, drowsy, sleeping, hyperventilating, photic stimulation, eyes open and eyes closed.

[l]Sleeping, hyperventilating, photic stimulation and nasopharyngeal leads.

some considerations should be taken into account. The studies of Sar et al. (2001, 2007b) only inform on steady state cerebral blood flow changes, i.e., a task-independent assessment of blood flow changes. In the study of Vermetten et al. (2006), the validity of the decrease in hippocampal and amygdalar volumes, when incorporating the age difference between DID subjects and the control group, has been discussed (Smeets, Jelicic, & Merckelbach, 2006; Vermetten, 2006). The PET study (Reinders, 2004; Reinders et al., 2003, 2006) applied a symptom provocation paradigm including a neutral and a trauma-related autobiographical stimulus script, which were presented to two DIS, i.e., a NIS and TIS. This study represents a two-by-two factorial design (Friston, Price, Buechel, & Frackowiak, 1997; Price, Moore, & Friston, 1997) and allows the assessment of various effects, namely main effects (both comprising two levels: NIS and TIS), interaction effects, several possibilities for simple subtraction analysis and conjunction analysis (see Figure 6.1 in Reinders, 2004). This study was set up as a within-subject study, where subjects served as their own control. Regional cerebral

TABLE 2
Brain regions found to be implicated in DID

Technique	References	Frontal	Parietal	Occipital	Temporal	Amygdala	Hippocampus
sMRI	Vermetten et al. (2006)[a]	x	x	x	x	R[b]+L	R[b]+L[b]
	Tsai et al. (1999)[c]	x	x	x	x	x	R+L
fMRI	Tsai et al. (1999)	–	–	–	R	–	R+L
PET	Reinders et al. (2006)[d]	R+L	R+L	R+L	R+L	R+L	R+L[e]
	Reinders et al. (2003)[d]	R+L	R+L	R+L	–	–	–
	Mathew et al. (1985)	–	–	–	R[f]	x	x
SPECT	Sar et al. (2007b)	R+L	–	R+L	–	x	x
	Sheehan et al. (2006)	–	–	–	R+L[f]	–	–
	Sar et al. (2001)	R+L	–	–	L	x	x
	Saxe et al. (1992)	~	~	~	L	~	~
ERP	Allen et al. (2000)	–	L+M[g]	–	–	x	x
EEG	Lapointe et al. (2006)	R	–	R+L	R+L	x	x
	Hughes et al. (1990)	–	R	R+L	R+L	x	x
	Coons et al. (1982)	–	R+L[h]	R[i]	L	x	x
	Mesulam et al. (1981)	–	–	–	R+L[f]	x	x

This table lists the location of statistically significant findings as reported in the neuroimaging studies listed in Table 1. As can be seen in Table 1 any convergence in neuroimaging studies in DID is missing. Furthermore, the spatial resolution of SPECT and EEG/ERP is poor. Therefore, this table only indicates whether or not a finding was reported in the listed brain site/lobe, i.e., it has a low spatial resolution reporting to conform to the low spatial resolution of SPECT and EEG/ERP. Listings in this table are independent of dissociative identity state, the task or paradigm under investigation, the comparison of task conditions, activation/deactivation, etc. It states only whether or not a region was found to be implicated in DID in the specific reference.

sMRI, structural magnetic resonance imaging; fMRI, functional magnetic resonance imaging; PET, positron emission tomography; SPECT, single photon emission computed tomography; ERP, event-related potential; EEG, electroencephalogram; –, was a region of interest but no significant results; x, not a region of interest; ~, unknown whether or not it was a region of interest; R, Right hemisphere; L, Left hemisphere; M, Midline.

[a]Manual tracing of hippocampus and amygdala.
[b]Not significant after adjustment for age (see Smeets et al., 2006 and Vermetten et al., 2006).
[c]Manual tracing of hippocampus.
[d]Talairach and Tournoux (1998) coordinates and Brodmann's areas in the original publication.
[e]Both R and L reported as parahippocampal instead of hippocampal.
[f]No statistical values available.
[g]Left posterior-parietal and posterior-parietal midline.
[h]Right centro-parietal + left parasagital.
[i]Right parieto-occipital.

blood flow (rCBF) data revealed different neural networks to be associated with the differential processing of neutral and trauma-related memory script between NIS and TIS. The results of this study confirm inter-identity amnesia between two identity states and support the traumagenic position.

However, one could argue that the PET results might vanish when correcting for the effect of simulation or that the autobiographical memories are (recovered) false memories, as both have been postulated by holders of the iatrogenic position. The transfer of newly learned information between identity states (i.e., testing inter-identity amnesia) was tested using psychological tests (Elzinga, Phaf, Ardon, & van Dyck, 2003; Huntjens, 2003; Huntjens et al., 2002, 2005a, 2006; Huntjens, Postma, Peters, Woertman, & van der Hart, 2003; Huntjens, Postma, Woertman, van der Hart &

Peters, 2005b). While Huntjens (2003) did find an inter-identity amnesia effect for genuine DID subjects, it was only partial and did not significantly differ from that of DID simulating control subjects, demonstrating the importance of including simulating subjects. Interestingly, as the studies of Huntjens (2003) speak against a complete state-dependent separation of memories, they might shed some light on the second and third categories of legal complications (see above and Merckelbach et al., 2002).

Besides (inter-identity) amnesia, another aspect of memory functioning is the involvement of false memories, which plays a key role in DID according to holders of the iatrogenic position. Exploring the neural correlates of false memories using functional neuroimaging is pioneering research. Although the first studies with healthy controls are appearing in the literature (Moritz, Glascher, Sommer, Büchel,

& Braus, 2006), the application to psychiatric disorders like PTSD or DID is, as far as the author is aware of, nonexistent.

Imaging deception

Pseudogenic DID involves the intention to deceive others, e.g., in the forensic setting the judge and jury, through actively malingering by simulating the disorder. Abe et al. (2006; Abe, Suzuki, Mori, Itoh, & Fujii, 2007) found, using PET, that the deception of an interrogator by making untruthful responses or telling lies involves the activation of the prefrontal cortex, anterior cingulate cortex and the amygdala. These findings are supported by pioneering fMRI studies (Kozel et al., 2005; Mohamed et al., 2006). The application of neuroimaging to investigate whether or not DID is pseudogenic looks promising on the basis of these studies. As these studies are performed in a laboratory setting with a group study of normal controls, it can be assumed that they underestimate the effects of malingering when it comes to real life and court cases. It can be expected that the emotional response and the effort to malinger in pseudogenic DID subject is greater. Both the PET and fMRI studies are pioneering multi-subject studies, which complicates the extension to a single-subject to warrant use in court (Appelbaum, 2007). Fast event-related and/or real-time fMRI on the other hand offer more potential (Haynes & Rees, 2006; Laconte, Peltier, & Hu, 2007; Langleben et al., 2005) as they can be applied in single subjects.

So, even though the developments in the field of brain imaging are interesting and promising, judicial decision-making in court cases concerning DID should incorporate the power and limitations of these techniques. When reading literature in this field, specific attention should be drawn to the possibilities *and* pitfalls of study designs in this type of published empirical research. Besides considering scientific literature during judiciary decision-making, judges or juries might be tempted to have the DID condition of the accused verified via one of the neuroimaging techniques mentioned above. However, given their limitations, such single-subject studies should strongly be discouraged. The currently available information (see Table 2) is obtained from poorly comparable brain imaging studies, which often lack methodological rigor (e.g., Tsai et al., 1999), and is therefore unsuitable for use in such delicate situations as legal proceedings (Kulynych, 1996; Merckelbach et al., 2002; Patel, Meltzer, & Mayberg, 2007). Imaging may be a promising tool to explore pathogenetic mechanisms in DID but at present the literature is very limited, without a single independently confirmed finding. Therefore, I strongly recommend the establishment of a consensus on distinguishing paradigms, acknowledged by both holders of the traumagenic and iatrogenic position, which are supported by a large number of well-controlled multi-subject studies, before considering neuroimaging tools as objective measures in court. As such studies and evaluations will take a significant amount of time, the availability of neuroimaging as a forensic neurobiological evaluative tool to forensic professionals will not be as soon as hoped for by Frankel and Dalenberg (2006, pp. 174–175). In addition, it will always remain difficult to compare a specific (psychopathological) case to such a database. Testing for pseudogenic DID, through accessing deception, needs the same caution as more information about the reliability and validity is required (Appelbaum, 2007).

CLOSING CONSIDERATIONS

As DID remains a popular topic among the general public it can be expected that (pseudogenic) DID will continue to play a role in court cases. Therefore, one of the main issues to be resolved around DID in forensic psychiatry is the etiology of the subject's DID. How can it be determined whether the origin of the subject's DID is traumagenic, iatrogenic or pseudogenic? Is the disorder genuine, subconsciously simulated or consciously malingered? And if the subject is subconsciously simulating DID due to iatrogenic intervention, who is the one to blame: subject or therapist?

Neuroimaging has been shown to be a powerful tool (Reinders et al., 2003, 2006; Sar et al., 2001, 2007b; Vermetten et al., 2006), revealing that genuine dissociative identity disorder is related to deviant amygdalar and hippocampal volumes and is characterized by at least two types of dissociative identity states: (i) dissociative identity states that process trauma-related memory as if they pertained to neutral memories, i.e., NIS, and consequently do not show differences in rCBF when listening to a trauma-related or neutral memory, and (ii) dissociative identity states fixated on, i.e., with access and responses to, traumatic memories do show differences in rCBF when listening to a trauma-related memory as compared to a neutral memory. Unfortunately, these studies do not inform us about the different neural correlates in relation to a traumagenic, iatrogenic and/or pseudogenic etiology of DID.

With the number of forensic cases involving DID increasing dramatically (Frankel & Dalenberg, 2006), convergence on diagnosing DID and distinguishing on the basis of etiology is needed. However, as long as there are no widely accepted methods and designs for (functional) neuroimaging case studies, for determining the neural correlates of the etiology of DID, severe caution is needed with regard to its use in court. Nevertheless, objectively measuring brain activation and structural patterns is one of the most promising directions for future scientific research. Such objective measures should be preferred over subjective measures as evidence in the courtroom, since the latter have been shown to be incapable of successfully identifying between genuine DID subjects and well-instructed simulating controls (Brand et al., 2006). Considering the rarity of empirical research conforming to methodological rigor, ethics and justice dictate that we keep an open mind and to consider all three possibilities, i.e., traumagenic, iatrogenic or pseudogenic, as the etiology of DID instead of choosing a position on the basis of emotion or prejudice.

REFERENCES

Abe, N., Suzuki, M., Tsukiura, T., Mori, E., Yamaguchi, K., Itoh, M., & Fujii, T. (2006). Dissociable roles of prefrontal and anterior cingulate cortices in deception. *Cerebral Cortex, 16*(2), 192–199.

Abe, N., Suzuki, M., Mori, E., Itoh, M., & Fujii, T. (2007). Deceiving others: Distinct neural responses of the prefrontal cortex and amygdala in simple fabrication and deception with social interactions. *Journal of Computational Neurology, 19*(2), 287–295.

Allen, J. J., & Movius, H. L. (2000). The objective assessment of amnesia in dissociative identity disorder using event-related potentials. *International Journal of Psychophysiology, 38*(1), 21–41.

American Psychiatric Association. (1994). *Diagnostic and Statistical Manual of Mental Disorders* (4th ed.). Washington, DC: American Psychiatric Press.

Appelbaum, P. S. (2007). Law & psychiatry: The new lie detectors: Neuroscience, deception, and the courts. *Psychiatric Services, 58*(4), 460–462.

Barlow, M. R. (2007). Researching dissociative identity disorder: Practical suggestions and ethical implications. *Journal of Trauma & Dissociation, 8*(1), 81–96.

Behnke, S. H. (1997). Confusion in the courtroom: How judges have assessed the criminal responsibility of individuals with multiple personality disorder. *International Journal of Law and Psychiatry, 20*(3), 293–310.

Bernstein, E. M., & Putnam, F. W. (1986). Development, reliability and validity of a dissociation scale. *The Journal of Nervous and Mental Disease, 174*, 727–735.

Bowman, E. S., Blix, S., & Coons, P. M. (1985). Multiple personality in adolescence: Relationship to incestual experiences. *Journal of the American Academy of Child Psychiatry, 24*(1), 109–114.

Brand, B. L., McNary, S. W., Loewenstein, R. J., Kolos, A. C., & Barr, S. R. (2006). Assessment of genuine and simulated dissociative identity disorder on the structured interview of reported symptoms. *Journal of Trauma & Dissociation, 7*(1), 63–85.

Brown, R. J. (2006). Different types of 'dissociation' have different psychological mechanisms. *Journal of Trauma & Dissociation, 7*(4), 7–28.

Brown, D., Scheflin, A., & Hammond, D. (1998). *Memory, trauma treatment, and the law.* New York: Norton.

Coons, P. M. (2005). Re: The persistence of folly: A critical examination of dissociative identity disorder. *Canadian Journal of Psychiatry, 50*(12), 813 (Comment).

Coons, P. M., Milstein, V., & Marley, C. (1982). EEG studies of two multiple personalities and a control. *Archives of General Psychiatry, 39*(7), 823–825.

Dalenberg, C. J., & Palesh, O. G. (2004). Relationship between child abuse history, trauma, and dissociation in Russian college students. *Child Abuse & Neglect, 28*(4), 461–474.

Deeley, P. Q. (2003). Social, cognitive, and neural constraints on subjectivity and agency: Implications for dissociative identity disorder. *Philosophy, Psychiatry and Psychology, 10*(2), 161–167.

DePrince, A. P., Allard, C. B., & Oh, H. (2004). What's in a name for memory errors? Implications and ethical issues arising from the use of the term 'False Memory' for errors in memory for details. *Ethics & Behavior, 14*(3), 201–233.

Dorahy, M. J. (2001). Dissociative identity disorder and memory dysfunction: The current state of experimental research and its future directions. *Clinical Psychology Review, 21*(5), 771–795.

Elzinga, B. M., Van Dyck, R., & Spinhoven, P. (1998). Three controversies about dissociative identity disorder. *Clinical Psychology and Psychotherapy, 5*(1), 13–23.

Elzinga, B. M., Phaf, R. H., Ardon, A. M., & van Dyck, R. (2003). Directed forgetting between, but not within, dissociative personality states. *Journal of Abnormal Psychology, 112*(2), 237–243.

Frankel, A. S., & Dalenberg, C. (2006). The forensic evaluation of dissociation and persons diagnosed with dissociative identity disorder: Searching for convergence. *Psychiatric Clinics of North America, 29*(1), 169–184 (Review).

Fraser, G. A. (2005). Re: The persistence of folly: A critical examination of dissociative identity disorder. *Canadian Journal of Psychiatry, 50*(12), 814 (Comment).

Friston, K. J., Price, C. J., Buechel, C., & Frackowiak, R. S. J. (1997). A taxonomy of study design. http://www.fil.ion.ucl.ac.uk/spm/course/notes97/Ch7.pdf, pp. 1–22.

Geraerts, E. (2006). *Remembrance of things Past: The cognitive psychology of remembering and forgetting trauma*. PhD thesis, University of Maastricht.

Geraerts, E., Smeets, E., Jelicic, M., van Heerden, J., & Merckelbach, H. (2005). Fantasy proneness, but not self-reported trauma is related to DRM performance of women reporting recovered memories of childhood sexual abuse. *Consciousness and Cognition, 14*(3), 602–612.

Gleaves, D. H. (1996). The sociocognitive model of dissociative identity disorder: A reexamination of the evidence. *Psychological Bulletin, 120*(1), 42–59 (Review).

Gleaves, D. H., May, M. C., & Cardena, E. (2001). An examination of the diagnostic validity of dissociative identity disorder. *Psychological Bulletin, 21*(4), 577–608 (Review).

Haynes, J. D., & Rees, G. (2006). Decoding mental states from brain activity in humans. *Nature Reviews Neuroscience, 7*(7), 523–534.

Hughes, J. R., Kuhlman, D. T., Fichtner, C. G., & Gruenfeld, M. J. (1990). Brain mapping in a case of multiple personality. *Clinical Electroencephalography, 21*(4), 200–209.

Huntjens, R. J. C. (2003). *Apparent amnesia: Interidentity memory functioning in dissociative identity disorder*. PhD thesis, University of Utrecht.

Huntjens, R. J., Postma, A., Hamaker, E. L., Woertman, L., van der Hart, O., & Peters, M. (2002). Perceptual and conceptual priming in patients with dissociative identity disorder. *Memory & Cognition, 30*(7), 1033–1043.

Huntjens, R. J., Postma, A., Peters, M. L., Woertman, L., & van der Hart, O. (2003). Interidentity amnesia for neutral, episodic information in dissociative identity disorder. *Journal of Abnormal Psychology, 112*(2), 290–297.

Huntjens, R. J., Peters, M. L., Postma, A., Woertman, L., Effting, M., & van der Hart O. (2005a). Transfer of newly acquired stimulus valence between identities in dissociative identity disorder (DID). *Behaviour Research and Therapy, 43*(2), 243–255.

Huntjens, R. J., Postma, A., Woertman, L., van der Hart O., & Peters, M. L. (2005b). Procedural memory in dissociative identity disorder: When can inter-identity amnesia be truly established? *Consciousness and Cognition, 14*(2), 377–389.

Huntjens, R. J., Peters, M. L., Woertman, L., Bovenschen, L. M., Martin, R. C., & Postma, A. (2006). Inter-identity amnesia in dissociative identity disorder: A simulated memory impairment? *Psychological Medicine, 36*(6), 857–863.

Kozel, F. A., Johnson, K. A., Mu, Q., Grenesko, E. L., Laken, S. J., & George, M. S. (2005). Detecting deception using functional magnetic resonance imaging. *Biological Psychiatry, 58*(8), 605–613.

Kulynych, J. (1996). Brain, mind, and criminal behavior: Neuroimages as scientific evidence. *Jurometrics Journal, 36*(3), 235–244.

Laconte, S. M., Peltier, S. J., & Hu, X. P. (2007). Real-time fMRI using brain-state classification. *Human Brain Mapping, 28*(10), 1033–1044.

Laney, C., & Loftus, E. F. (2005). Traumatic memories are not necessarily accurate memories. *Canadian Journal of Psychiatry, 50*(13), 823–828 (Review).

Langleben, D. D., Loughead, J. W., Bilker, W. B., Ruparel, K., Childress, A. R., Busch, S. I., & Gur, R. C. (2005). Telling truth from lie in individual subjects with fast event-related fMRI. *Human Brain Mapping, 26*(4), 262–272.

Lapointe, A. R., Crayton, J. W., DeVito, R., Fichtner, C. G., & Konopka, L. M. (2006). Similar or disparate brain patterns? The intra-personal EEG variability of three women with multiple personality disorder. *Clinical EEG and Neuroscience, 37*(3), 235–242.

Mathew, R. J., Jack, R. A., & West, W. S. (1985). Regional cerebral blood flow in a patient with multiple personality. *American Journal of Psychiatry, 142*(4), 504–505.

McElroy, L. P. (1992). Early indicators of pathological dissociation in sexually abused children. *Child Abuse & Neglect, 16*, 833–846.

Merckelbach, H., & Muris, P. (2001). The causal link between self-reported trauma and dissociation: A critical review. *Behaviour Research and Therapy, 39*, 245–254 (Review).

Merckelbach, H., Horselenberg, R., & Muris, P. (2001). The creative experiences questionnaire (CEQ): A brief self-report measure of fantasy proneness. *Personality and Individual Differences, 31*, 987–995.

Merckelbach, H., Devilly, G. J., & Rassin, E. (2002). Alters in dissociative identity disorder. Metaphors or genuine entities? *Clinical Psychology Review, 22*(4), 481–497 (Review).

Merckelbach, H., Smeets, T., Geraerts, E., Jelicic, M., Bouwen, A., & Smeets, E. (2006). I haven't thought about this for years! Dating recent recalls of vivid memories. *Applied Cognitive Psychology, 20*, 33–42.

Mesulam, M. M. (1981). Dissociative states with abnormal temporal lobe EEG: Multiple personality and the illusion of possession. *Archives of Neurology, 38*(3), 176–181.

Mohamed, F. B., Faro, S. H., Gordon, N. J., Platek, S. M., Ahmad, H., & Williams, J. M. (2006). Brain mapping of deception and truth telling about an ecologically valid situation: Functional MR imaging and polygraph investigation-initial experience. *Radiology, 238*(2), 679–688.

Moritz, S., Glascher, J., Sommer, T., Büchel, C., & Braus, D. F. (2006). Neural correlates of memory confidence. *NeuroImage, 33*(4), 1188–1193.

Nijenhuis, E. R. S., Spinhoven, P., van Dyck, R., van der Hart, O., & Vanderlinden, J. (1998). Degree of somatoform and psychological dissociation in dissociative disorder is correlated with reported trauma. *Journal of Traumatic Stress, 11*(2), 711–730.

Nijenhuis, E. R. S., van der Hart, O., & Steele, K. (2002). The emerging psychobiology of trauma-related dissociation and dissociative disorders. In H. A. H. D'haenen, J. A. Den Boer, & P. Willner (Eds), *Biological psychiatry* (vol. 2, pp. 1079–1098). Chichester, UK: Wiley & Sons Ltd.

Patel, P., Meltzer, C. C., Mayberg, H. S., & Levine, K. (2007). The role of imaging in United States courtrooms. *Neuroimaging Clinics of North America, 17*(4), 557–567.

Piper, A., & Merskey, H. (2004a). The persistence of folly: A critical examination of dissociative identity disorder. Part I. The excesses of an improbable concept. *Canadian Journal of Psychiatry, 49*(9), 592–600 (Review).

Piper, A., & Merskey, H. (2004b). The persistence of folly: Critical examination of dissociative identity disorder. Part II. The defence and decline of multiple personality or dissociative identity disorder. *Canadian Journal of Psychiatry, 49*(10), 678–683 (Review).

Piper, A., & Merskey, H. (2005). The persistence of folly: A critical examination of dissociative identity disorder. *Canadian Journal of Psychiatry, 50*(12), 814 (Comment).

Pope, H. G., Barry, S., Bodkin, A., & Hudson, J. I. (2006). Tracking scientific interest in the dissociative disorders: A study of scientific publication output 1984–2003. *Psychotherapy and Psychosomatics, 75*, 19–24.

Price, C. J., Moore, C. J., & Friston, K. J. (1997). Subtractions, conjunctions, and interactions in experimental design of activation studies. *Human Brain Mapping, 5*(4), 264–272.

Reinders, A. A. T. S. (2004). Psycho-biological characteristics of dissociative identity disorder: rCBF, physiologic, and subjective findings from a symptom provocation study. In *From methods to meaning in functional neuroimaging* (pp. 63–93). University Library Groningen, Groningen. PhD thesis.

Reinders, A. A. T. S., Nijenhuis, E. R. S., Paans, A. M. J., Korf, J., Willemsen, A. T. M., & den Boer, J. A. (2003). One brain, two selves. *NeuroImage, 20*(4), 2119–2125.

Reinders, A. A. T. S., Nijenhuis, E. R. S., Quak, J., Korf, J., Haaksma, J., Paans, A. M. J., Willemsen, A. T. M., & den Boer, J. A. (2006). Psychobiological characteristics of dissociative identity disorder: A symptom provocation study. *Biological Psychiatry, 60*(7), 730–740.

Rieber, R. W. (1999). Hypnosis, false memory and multiple personality: A trinity of affinity. *History of Psychiatry, 10*(37), 3–11.

Rogers, R. (1997). *Clinical assessment of malingering and deception* (2nd ed.). New York: Guilford Press.

Ross, C. A., Norton, G. R., & Wozney, K. (1989). Multiple personality disorder: An analysis of 236 cases. *Canadian Journal of Psychiatry, 34*(5), 413–418.

Sar, V. (2005). Re: The persistence of folly: A critical examination of dissociative identity disorder. What are Dr Piper and Dr Merskey trying to do? *Canadian Journal of Psychiatry, 50*(12), 813 (Comment).

Sar, V., Unal, S. N., Kiziltan, E., Kundaci, T., & Ozturk, E. (2001). HMPAO SPECT study of regional cerebral blood flow in dissociative identity disorder. *Journal of Trauma & Dissociation, 2*(2), 5–25.

Sar, V., & Akyuzi, G. Dogan, O. (2007a). Prevalence of dissociative disorders among women in the general population. *Psychiatry Research, 149*(1–3) 169–76.

Sar, V., Unal, S. N., & Ozturk, E. (2007b). Frontal and occipital perfusion changes in dissociative identity disorder. *Psychiatry Research: Neuroimaging, 156*(3), 217–223.

Saxe, G. N., Vasile, R. G., Hill, T. C., Bloomingdale, K., & Van der Kolk, B. A. (1992). SPECT imaging and multiple personality disorder. *The Journal of Nervous and Mental Disease, 180*(10), 662–663.

Schreiber, F. (1973). *Sybil.* New York: Warner Communications.

Sheehan, W., Thurber, S., & Sewall, B. (2006). Dissociative identity disorder and temporal lobe involvement: Replication and a cautionary note. *The Royal Australian and New Zealand College of Psychiatrists, 40*(4), 374–375.

Smeets, T., Jelicic, M., & Merckelbach, H. (2006). Reduced hippocampal and amygdalar volume in dissociative identity disorder: Not such clear evidence. *American Journal of Psychiatry, 163*(9), 1643 (Comment).

Spanos, N. P. (1994). Multiple identity enactments and multiple personality disorder: A sociocognitive perspective. *Psychological Bulletin, 116*(1), 143–165 (Review).

Steele, K., van der Hart, O., & Nijenhuis, E. (2001). Dependency in the treatment of complex posttraumatic stress disorder and dissociative disorders. *Journal of Trauma & Dissociation, 2*(4), 79–116.

Steinberg, M. (1993). *Structured clinical interview for DSM-IV dissociative disorders (SCID-D).* Washington, DC: American Psychiatric Press.

Stern, C. E., Corkin, S., Gonzalez, R. G., Guimaraes, A. R., Baker, J. R., Jennings, P. J., Carr, C. A., Sugiura, R. M., Vedantham, V., & Rosen, B. R. (1996). The hippocampal formation participates in novel picture encoding: Evidence from functional magnetic resonance imaging. *Proceedings of the National Academy of Sciences of the USA, 93*, 8860–8865.

Stickley, T., & Nickeas, R. (2006). Becoming one person: Living with dissociative identity disorder. *Journal of Psychiatric and Mental Health Nursing, 13*(2), 180–187 (Review).

Talairach, J., & Tournoux, P. (1988). *Co-planar stereotaxic atlas of the human brain.* Stuttgart: Thieme Verlag.

Tsai, G. E., Condie, D., Wu, M. T., & Chang, I. W. (1999). Functional magnetic resonance imaging of personality switches in a woman with dissociative identity disorder. *Harvard Review of Psychiatry, 7*(2), 119–122.

Van der Kolk, B. A., & Fisler, R. (1995). Dissociation and the fragmentary nature of traumatic memories: Overview and exploratory study. *Journal of Traumatic Stress, 8*, 505–525.

Vermetten, E. (2006). Reduced hippocampal and amygdalar volume in dissociative identity disorder: Not such clear evidence. *American Journal of Psychiatry, 163*(9), 1643–1644 (Reply).

Vermetten, E., Schmahl, C., Lindner, S., Loewenstein, R. J., & Bremner, J. D. (2006). Hippocampal and amygdalar volumes in dissociative identity disorder. *American Journal of Psychiatry, 163*(4), 630–636.

Xiao, Z., Yan, H., Wang, Z., Zou, Z., Xu, Y., Chen, J., Zhang, H., Ross, C. A., & Keyes, B. B. (2006). Trauma and dissociation in China. *American Journal of Psychiatry, 163*(8), 1388–1391.

Zlotnick, C., Shea, M. T., Pearlstein, T., Begin, A., Simpson, E., & Costello, E. (1996). Differences in dissociative experiences between survivors of childhood incest and survivors of assault in adulthood. *The Journal of Nervous and Mental Disease, 184*(1), 52–54.

NEUROCASE
2008, 14 (1), 54–58

Developing a Neuropsychiatric Functional Brain Imaging Test

F. Andrew Kozel and Madhukar H. Trivedi

Department of Psychiatry, University of Texas Southwestern Medical Center, Dallas, TX, USA

A number of critical issues must be addressed in order to develop and properly apply a functional brain imaging test. Diagnostic tests involve making a judgment for a single person. As a result, functional brain imaging tests must also be evaluated at the individual level. The population examined in determining the evidence for the accuracy of the test and the specific question being tested should be clearly described so that the test can be applied appropriately. The accuracy of the test must also be established in order to know the degree of confidence to accord a result. Incorporating what has been learned with medical diagnostic test development will enable legitimate and significant neuropsychiatric functional brain imaging tests to be developed in the future.

Keywords: Brain imaging; Diagnostic; Test; Neuropsychiatry; Detecting deception; Functional MRI.

Advancing a neuroscience tool like neuroimaging to detect a brain function that will be used to understand some aspect of an individual's behavior should be conceptualized as developing a medical diagnostic test. As such, the development and understanding of the neuroimaging test should be informed by what has been learned from diagnostic testing research in other areas of study. Several critical issues will be presented which must be considered when developing and applying a neuropsychiatric functional brain imaging diagnostic test. Certain issues are unique to developing a neuroimaging diagnostic test in neuropsychiatry while others are common to any diagnostic test development. In addition to test development, the importance of the proper application of a test will be discussed.

There is a growing interest and literature in the science of diagnostic test development (Bossuyt et al., 2003a; Boutros, Fraenkel, & Feingold, 2005; Buntinx & Knottnerus, 2006; Knottnerus, 2002; Knottnerus & Muris, 2003; Knottnerus, van Weel, & Muris, 2002; Ransohoff, 2002; Sackett & Haynes, 2002). The purpose of these efforts is to ensure that the information obtained from these tests can be appropriately interpreted and applied to the *individual* of interest. This purpose in turn leads to a fundamental issue which all tests share – i.e., is there adequate scientific support for the test *and* its intended use?

In order to make an informed decision regarding the utility of a test, we must understand the information obtained from studies and use it in an appropriate manner when translating scientific research to application. To state that a test has merit based solely on 'expert opinion' or 'years of experience' with no supporting scientific evidence

Thanks to Dr Steven J. Laken for review and comment on the manuscript. Dr Kozel is supported by a K23 from the National Institute of Mental Health (5K23MH070897). The content is solely the responsibility of the authors and does not necessarily represent the official views of the National Institute of Mental Health or the National Institutes of Health. Dr Kozel is listed as an inventor on patent applications being applied for by the Medical University of South Carolina for the fMRI Detection of Deception. Dr Kozel has also received research support from Cephos Corp. and the Defense Agency for Credibility Assessment, U. S. Department of Defense.

Address correspondence to F. Andrew Kozel, M.D., M.S.C.R., Assistant Professor, Department of Psychiatry, UT Southwestern Medical Center, 5323 Harry Hines Blvd, Dallas, TX 75390-9119, USA (E-mail: Andrew.Kozel@utsouthwestern.edu).

DOI: 10.1080/13554790701881731

is not only deceitful but dangerous. There are numerous possibilities for doing harm by arriving at erroneous conclusions. These include: acts of inappropriate attribution such as declaring someone as not culpable for ones criminal behavior due to a perceived brain defect on a neuroimaging scan when the neuroimaging finding had nothing to do with causing the criminal behavior; acts of failure to properly ascribe cause to a brain pathology such as pronouncing someone as blameworthy because no brain defect could be determined on a brain scan when the individual was in fact quite impaired; and acts of incorrect judgments such as providing evidence to exonerate someone when the person is guilty or conversely, providing evidence to convict someone who is innocent. Although these and other potential pit-falls could exist if tests are not applied carefully, a number of potential advantages exist that advanced neuroimaging technologies have to offer in order to assist with decision making. We are beginning to measure the function of the brain in ways which will help us to better identify, understand, and potentially treat neuropsychiatric conditions such as depression, mania, psychosis, as well as evaluate conditions like sociopathy (Dougherty & Rauch, 2007; Gur, Keshavan, & Lawrie, 2007; Kruesi & Casanova, 2006; Kruesi, Casanova, Mannheim, & Johnson-Bilder, 2004; Phillips & Vieta, 2007; Raine, Lencz, Bihrle, LaCasse, & Colletti, 2000). In addition, the ability to use functional imaging at the individual level is being explored in recent studies to develop paradigms that asses the function of cingulo-frontal-parietal circuits (Bush & Shin, 2006) and to more accurately detect deception (Davatzikos et al., 2005; Johnson, George, & Kozel, in press; Kozel et al., 2005; Langleben et al., 2005). Ultimately, the ability to better understand the relationship between certain brain abnormalities and culpability would provide much needed help for the legal system.

Whether these tests are ultimately used probably will not be decided by science. The decision, however, to use a test should be based on appropriate and meaningful scientific information. Providing information regarding the strengths, weaknesses, and what is *not known* about a given test is critical to the evaluation of such tests in particular situations. In addition, understanding whether the test is assessing a 'non-activated' (sometimes referred to as 'resting state' – although the living brain never rests) brain state or a specifically 'activated' brain state is essential to interpreting the meaning of a neuroimaging test. Measuring a non-activated brain state provides a general measure of brain function and does not provide any specific information regarding a particular task or behavior. Conversely, measuring brain activity during a task enables the brain function to be assessed during that particular task. Each type of testing has the potential for usefulness. Both non-activated cardiac (EKG) and activated cardiac tests (stress test) play an important role in the evaluation of the heart, but one must appreciate what is being evaluated.

A number of challenges exist for developing a neuropsychiatric functional brain imaging test. Several of these challenges are related to specific aspects of functional neuroimaging and others are common to the development of all new diagnostic tests. Specific to this technology, although there is an extensive literature of using functional neuroimaging to better understand brain based disorders, the majority of the reports involve group averaged data and successful replication is not the norm. Additional concerns include the facts that functional neuroimaging measures only correlates of brain function versus the basis of brain function (i.e., causation) and the lack of a neuropathological reference standard in most neuropsychiatric disorders makes developing a neuropsychiatric brain imaging test difficult. Common to the development of all new diagnostic tests; attempting to control for potential bias in the results, knowing the true predictive value of the test, and understanding how the test will work in different groups of people requires carefully designed studies to address these concerns.

A challenge specific to developing a neuroimaging diagnostic test is the fact that the majority of functional neuroimaging research has focused on group analysis (summing together the data from the individual participants). The focus on group analysis is largely driven by the low signal to noise ratio that is present in much of functional neuroimaging. For a neuroimaging diagnostic test to be useful, however, it must provide reliable and valid results for an individual. Group averaged data does not provide any information specific to an individual. Thus, using group averaged data to support any conclusions made on an individual's scan is unwarranted and disingenuous. Group analysis can, however, play an important role in the initial development of a diagnostic test. If differences in appropriately matched groups or conditions occur, this provides support for further investigation at the individual level.

Another specific challenge is the nature of current functional neuroimaging. Presently, functional neuroimaging studies measure aspects of brain function that are correlated to brain activity and/or the summation of large portions of brain. Thus, we are not measuring directly the brain function that we are most interested in assessing. This does not, however, mean that neuroimaging cannot be developed as a useful diagnostic test. There are numerous tests in medicine that have made a significant impact on clinical care, but are correlational in nature. Virtually all cancer screening tests are correlational (e.g., ossification on mammograms, high PSA tests, occult blood results, and lung nodules on X-ray). Another common example in medicine is to test for antibodies or an exaggerated immune response to an infectious agent in order to determine if the infection is present. One method to test for tuberculosis (*Mycobacterium tuberculosis*) involves evaluating the response to injected purified protein derivatives of *M. tuberculosis* (PPD) (Wallach, 1996). The antibodies are not causing the pathology, but are highly correlated with the infection. What must be taken into account, however, is that factors (disease, medications, etc.) which impact on the correlation of the measured information (e.g., degree of induration at injection site) with the information of interest (e.g., tuberculosis) could impact the worthiness of the test (Wallach, 1996). In a similar manner, factors (e.g., age, medications, medical conditions) which can alter the relationship between brain function and the measured signal (e.g., BOLD signal) need to be understood or controlled for when developing a functional neuroimaging test.

Because the nature of various factors on the correlation between measured signal and brain function is not well understood, a critical point is that the person of interest being tested is similar to those who were tested in the research. For example, if the research supporting a test was based on healthy, un-medicated, young graduate students, understanding how to interpret the result of this test on a 70-year-old person with cognitive impairment who is taking many medications is not possible. Importantly, to say that the test *will not work* in this demographic is just as unreasonable. Without data one way or the other, one can only say that the results of the tests are not able to be interpreted with the available information.

In addition to understanding how different demographics may effect the test results, knowing the predictive value associated with that test in a given population is of critical importance (Ransohoff, 2002). In other words, if the test is positive, what is the probability that the person has the condition. Similarly, if the test is negative, what is the probability that the person does not have the condition. Just as in other areas of science, if you do not know the error surrounding a result, you cannot understand the determined value.

There are several difficulties to determining the true predictive value of a brain imaging diagnostic test in neuropsychiatry. A principle problem is the lack of an ideal reference standard (a.k.a. Gold Standard) against which to assess a diagnostic test. Although some conditions such as Alzheimer's disease can be confirmed at autopsy (Rowland, 1995), most neuropsychiatric conditions such as depression, psychosis, or sociopathy cannot be biologically diagnosed – even at autopsy. Thus, the diagnostic test must be compared with other ways of coming to the diagnosis such as consensus agreement of clinical features, longitudinal course of condition, etc. (Knottnerus & Muris, 2003). These methods have obvious shortcomings which must be recognized when interpreting scientific results from studies using them as the reference standard.

Another major challenge in determining the true predictive value of any diagnostic test is the impact of bias on the reported results concerning that test (Knottnerus et al., 2002). In other words, something about the study design may have been more responsible for the test results than the condition of interest. There are a number of sources of potential bias. Ways to help combat the potential for bias are similar to the methodologies used in clinical trials to reduce bias. In determining if a neuroimaging test can differentiate whether a condition is present (e.g., depressed versus nondepressed), subjects from the two groups being compared should be drawn from similar populations (e.g., healthy, young adults with depression seeking care at a general primary care clinic *versus* healthy, young adults without depression seeking care at a general primary care clinic). If the participants being compared are not from similar populations, then one cannot be sure that the difference is due to the condition (e.g., depression) or to a difference in the populations studied. For example, comparing the neuroimaging findings from depressed, older, medically ill patients with healthy, young adults without depression can result in seeing differences that are due to age or medical status and not depression.

An additional source of important bias is if the investigator uses prior knowledge of a condition (e.g., participant was taking an antipsychotic) to help with the test procedure (e.g., testing for schizophrenia). Keeping investigators who perform the various measures masked (blind) to the condition being investigated helps ensure that the results are not impacted by prior knowledge (Sackett & Haynes, 2002). Also, automated measures that do not require decisions to be made by the investigator have the advantage of both reducing this risk as well as reducing the variability that can exist between investigators in carrying out the test. In order to interpret the research studies regarding diagnostic tests properly, one must consider how bias could have impacted the reported results in the studies.

As brain imaging diagnostic tests are developed, in addition to properly understanding the test, the test must be properly applied (i.e., the test must be used in the manner in which it was tested). As an example, suppose a fictional test was evaluated and revealed differences in brain function between individuals in prison versus people who are not in prison. Even assuming the studies were designed correctly, evaluated participants at the individual level, included appropriate population-based samples, and the results were adequately replicated – the test could still only be used to determine if the individual being tested had a brain function consistent with other prisoners. The test could not be used to determine what role that difference in brain function played in committing a crime, and it could not be used to determine if the individual committed the crime in question. There would be no data supporting causality – only a correlational measure with a sample of the population who were convicted of committing a crime.

In conclusion, the development of a neuropsychiatric brain imaging diagnostic test shares the same requirements of all good science (Bossuyt et al., 2003b). The purpose of the diagnostic test should be clearly stated as well as what the diagnostic test will assess. The manner in which the test has been examined and study populations evaluated are important to determine how or if the study's results can be applied to the person being tested. Only results based on measurement at the individual level can be used to support the utility of a test of a single person. The accuracy of the diagnostic ability of the test is critical to understanding how likely the result truly represents the condition of interest. Because of the potential for random

effects, unintended confounds, and biases to impact the results of studies, replication with independent samples is required to have confidence in any results. This is especially true if data from the participants were used to develop the methods for the analysis. The diagnostic test also should be evaluated across multiple setting and populations to better understand the strengths and limitations of the test. As the ability to develop neuroimaging tests advances, a fundamental principle is that a test can only be used to inform a situation upon which it has previously been scientifically evaluated. Using what has been learned in medical diagnostic test assessment will be important in developing legitimate and valuable neuropsychiatric brain imaging tests in the future.

REFERENCES

Bossuyt, P. M., Reitsma, J. B., Bruns, D. E., Gatsonis, C. A., Glasziou, P. P., Irwig, L. M., et al. (2003a). Towards complete and accurate reporting of studies of diagnostic accuracy: The STARD initiative. Standards for Reporting of Diagnostic Accuracy. *Clinical Chemistry, 49*, 1–6.

Bossuyt, P. M., Reitsma, J. B., Bruns, D. E., Gatsonis, C. A., Glasziou, P. P., Irwig, L. M., et al. (2003b). The STARD statement for reporting studies of diagnostic accuracy: Explanation and elaboration. *Clinical Chemistry, 49*, 7–18.

Boutros, N., Fraenkel, L., & Feingold, A. (2005). A four-step approach for developing diagnostic tests in psychiatry: EEG in ADHD as a test case. *Journal of Neuropsychiatry and Clinical Neurosciences, 17*, 455–464.

Buntinx, F., & Knottnerus, J. A. (2006). Are we at the start of a new era in diagnostic research? *Journal of Clinical Epidemiology, 59*, 325–326.

Bush, G., & Shin, L. M. (2006). The Multi-Source Interference Task: An fMRI task that reliably activates the cingulo-frontal-parietal cognitive/attention network. *Natural Protocols, 1*, 308–313.

Davatzikos, C., Ruparel, K., Fan, Y., Shen, D. G., Acharyya, M., Loughead, J. W., et al. (2005). Classifying spatial patterns of brain activity with machine learning methods: Application to lie detection. *Neuroimage, 28*, 663–668.

Dougherty, D. D., & Rauch, S. L. (2007). Brain correlates of antidepressant treatment outcome from neuroimaging studies in depression. *Psychiatric Clinics of North America, 30*, 91–103.

Gur, R. E., Keshavan, M. S., & Lawrie, S. M. (2007). Deconstructing psychosis with human brain imaging. *Schizophrenia Bulletin, 33*, 921–931.

Johnson, K. A., George, M. S., & Kozel, F. A. (in press). Detecting deception using functional magnetic resonance imaging. *Directions in Psychiatry*.

Knottnerus, J. A. (2002). *The Evidence Base of Clinical Diagnosis*. London: British Medical Journal Books.

Knottnerus, J. A., & Muris, J. W. (2003). Assessment of the accuracy of diagnostic tests: The cross-sectional study. *Journal of Clinical Epidemiology, 56,* 1118–1128.

Knottnerus, J. A., van Weel, C., & Muris, J. W. (2002). Evaluation of diagnostic procedures. *British Medical Journal, 324,* 477–480.

Kozel, F. A., Johnson, K. A., Mu, Q., Grenesko, E. L., Laken, S. J., & George, M. S. (2005). Detecting deception using functional magnetic resonance imaging. *Biological Psychiatry, 58,* 605–613.

Kruesi, M. J., & Casanova, M. F. (2006). White matter in liars. *British Journal of Psychiatry, 188,* 293–294; author reply 294.

Kruesi, M. J., Casanova, M. F., Mannheim, G., & Johnson-Bilder, A. (2004) Reduced temporal lobe volume in early onset conduct disorder. *Psychiatry Research, 132,* 1–11.

Langleben, D. D., Loughead, J. W., Bilker, W. B., Ruparel, K., Childress, A. R., Busch, S. I., et al. (2005). Telling truth from lie in individual subjects with fast event-related fMRI. *Human Brain Mapping, 26,* 262–272.

Phillips, M. L., & Vieta, E. (2007). Identifying functional neuroimaging biomarkers of bipolar disorder: Toward DSM-V. *Schizophrenia Bulletin, 33,* 893–904.

Raine, A., Lencz, T., Bihrle, S., LaCasse, L., & Colletti, P. (2000). Reduced prefrontal gray matter volume and reduced autonomic activity in antisocial personality disorder. *Archives of General Psychiatry, 57,* 119–127.

Ransohoff, D. F. (2002). Challenges and opportunities in evaluating diagnostic tests. *Journal of Clinical Epidemiology, 55,* 1178–1182.

Rowland, L. P. (Ed.) (1995). *Merritt's Textbook of Neurology* (9th ed.), Baltimore, MD: Williams & Wilkins.

Sackett, D. L., & Haynes, R. B. (2002). The architecture of diagnostic research. *British Medical Journal, 324,* 539–541.

Wallach, J. (1996). *Interpretation of Diagnostic Tests* (6th ed.), Boston, MA: Little, Brown and Company.

NEUROCASE
2008, 14 (1), 59–67

fMRI investigation of the cognitive structure of the Concealed Information Test

J. G. Hakun, D. Seelig, K. Ruparel, J. W. Loughead, E. Busch, R. C. Gur, and D. D. Langleben

Department of Psychiatry, University of Pennsylvania and the Veterans Administration Medical Center, Philadelphia, PA 19104, USA

We studied the cognitive basis of the functional magnetic resonance imaging (fMRI) pattern of deception in three participants performing the Concealed Information Test (CIT). In all participants, the prefrontoparietal lie activation was similar to the pattern derived from the meta-analysis ($N = 40$) of our previously reported fMRI CIT studies and was unchanged when the lie response was replaced with passive viewing of the target items. When lies were replaced with irrelevant responses, only the left inferior gyrus activation was common to all subjects. This study presents a systematic strategy for testing the cognitive basis of deception models, and a qualitative approach to single-subject truth-verification fMRI tests.

Keywords: fMRI; Deception; Lie-detection; Guilty knowledge test; Concealed information test; Control question test; GKT; CIT; CQT; Left inferior frontal gyrus; Ventro-medial Prefrontal Cortex; Meta-analysis.

INTRODUCTION

There are two ways to determine the truth: detect the truth itself or detect the lie and infer the opposite. Physiological measurements during a formal forced-choice or free-response query are the basis of most methods used to determine the subjective truth. The query protocols are relatively independent from the method used to collect physiological data, making the body of knowledge about these protocols acquired with polygraph measures, relevant for the newer measures such as fMRI. Most formal query paradigms used for truth determination with physiological markers, belong to one of two formats: the Control Question Test (CQT), directed at identifying the lie by comparing possible lie with known truth, and the Concealed Information Test (CIT), also known as the Guilty Knowledge Test (GKT), focused at identifying the physiological markers of 'concealed' knowledge directly. In its classic form, the CIT involves negative answers to a series of questions, some of which are related to the topic of the interrogation. The CIT does not discriminate between physiological response due to lying and orienting or other reasons, such as simple recognition of a stimulus (Lykken, 1991) and is thereby not considered a 'lie-detection test' in the strict sense of the word (Verschuere, Crombez, De Clercq, & Koster, 2004). The extent to which nonspecific physiological responses are controlled in the CQT depends on the closeness of match between the presumed lie and control items. Thus, the CQT and CIT are extremes on a continuum, as evidenced by the introduction of hybrid deception-generating paradigms (Furedy, Gigliotti, & Ben-Shakhar, 1994). One paradigm that our group has recently used to successfully discriminate between lie and truth with fMRI in a laboratory setting,

This work was supported in part by an unrestricted grant from No Lie MRI, Inc.

Address correspondence to D. D. Langleben, Treatment Research Center, University of Pennsylvania, 3900 Chestnut St., Philadelphia, PA 19104-6178, USA. E-mail: langlebe@mail.med.upenn.edu

http://www.psypress.com/neurocase

DOI: 10.1080/13554790801992792

has been referred to as a 'modified GKT (CIT)' (Langleben et al., 2005). This model incorporated control (i.e., known truth) items that were used to compare fMRI signal during lie and truth. The early fMRI studies of deception used the overlapping pre-fronto-parietal (PFP) lie pattern of the CIT and CQT type paradigms to postulate that response inhibition and behavioral control were the key cognitive components of deception (Langleben et al., 2002; Spence et al., 2001, 2004). Though assuming that the PFP pattern is specific to at least some forms of deception is tempting, the very structure of these tasks suggests that they may engage systems that are also involved in other types of cognition and behavior such as working memory, exogenous and endogenous attention, and behavioral and cognitive control. Dissociating the cognitive processes specific to lying from these parallel cognitive processes could enhance the accuracy of fMRI-based lie detection. The goal of this pilot study was to manipulate some of the key parameters of the CIT in order to provide an experimental framework for formally investigating basic cognitive operations such as attention, orienting, and working memory, all of which may contribute to the pattern of fMRI signal elicited by forced-choice deception paradigms.

Our working hypotheses were:

1. Brain response during deception elicited by the standard binary forced-choice CIT ('Stim Test') will be similar to the average deceptive response pattern observed in our prior fMRI CIT studies (Langleben et al., 2002, 2005).
2. The observed Stim Test pattern will remain unchanged after the following manipulations of the CIT:

   a. passive viewing of the CIT task stimuli, in the absence of a query or deceptive response;
   b. replacing deceptive and truthful responses with irrelevant (non-deceptive, evaluative) responses.

METHODS

Subjects

Subjects were three healthy, right-handed, English-speaking, college-educated females (16 years of education), 24, 24 and 25 years of age. To produce a regions-of-interest (ROI) template for qualitative ana-

lysis of the single-subject data, data from two previously reported fMRI CIT studies (Langleben et al., 2002, 2005) were subjected to meta-analysis. These data were acquired from 40 (12 female, 28 male) right-handed, English-speaking participants, 19–50 years of age. Each subjects' medical and psychiatric status was ascertained through a detailed assessment by a board-certified physician (D.D.L.). Substance abuse was excluded by a urine drug test. Candidates receiving prescription medications, those with a history of DSM IV Axis I psychiatric disorder, as well as those with any chronic medical illness or significant past trauma, were excluded. The study protocol was approved by the University of Pennsylvania Institutional Review Board.

Experimental procedure

Subjects participated in three consecutive tasks: a CIT in a format referred to by polygraph examiners as the 'Stim Test' (ST); (Matte, 1996; Elaad & Kleiner, 1986) and two manipulations of this task: Irrelevant Query CIT (IRQ); and Orienting CIT (OR). The ST was administered twice: before the scan session as a training session and during the scan session. The IRQ and OR were administered during the scan session only. After screening and informed consent, a designated team member met subjects, gave them the ST instructions, and conducted the ST. The subjects were then escorted to the scanner, where they were greeted by a different team member and performed the three tasks in the scanner, in a single fMRI session. During the scan session, the ST was performed first, and the order of the IRQ and OR was counterbalanced. Additional instructions specific to the IRQ and the OR were delivered via headphones prior to each task. The order of the fMRI tasks for Subject 1 and Subject 3 was ST, IRQ, and OR; and for Subject 2 the order was: ST, OR, IRQ.

Task instructions

The examiner asked each participant to pick a number from 3 through 8 (inclusive), write it down in secret on a separate sheet of paper, and place the paper in their pocket for the remainder of the study. Then the examiner presented each participant with the numbers 1 through 9 (inclusive) written on a sheet of paper. Subjects were instructed to deny having written the number they had in their pocket when asked about it and to tell the truth in response to questions about all other numbers.

After the questioning, the examiner instructed the subjects to adhere to the same instructions when questioned in the scanner.

Task design

Stim Test (ST)

The fMRI paradigm design of the ST was sparse event-related (Aguirre & D'Esposito, 1999; Dale, 1999). Stimuli were white numbers 1 through 9 (inclusive) presented on a black background, accompanied by a question: 'Do you have the number (X)?' Emulation of the green and blue response buttons of the fiber-optic response pad (fORP; Current Design, Philadelphia, PA) appeared on the bottom of the screen, with the words 'YES' and 'NO' above them, respectively. Stimuli classes were: Truth (numbers 3 through 8, inclusive, less the chosen number); Lie (the chosen number); and Control-Truth (numbers 1, 2, and 9).

Each stimulus, 1 through 9, was repeated 5 times throughout the experiment. Stimuli were presented for 3 s and separated by variable ISIs (10–16 s, mn = 13 s). The first presentation of each stimulus was in ascending numerical order, the second, third and fourth presentations of each number were in pseudorandom order, and the last presentation of each stimulus was in descending numerical order. This order approximated the polygraph ST format (Matte, 1996). The question appeared at the top of the screen, and stimuli were presented center-screen below the question (target area). During query trials, stimuli were presented in the target area; between each query trial (ISI) a fixation cross ('+') appeared in the target area. Subjects were asked to make a response to query trials using the fiber-optic response pad (fORP). Stimuli were presented and responses logged by Presentation® software (Version 0.70, www.neurolabs.com). The ST was 11 min 15 s long.

Irrelevant Query task (IRQ)

Stimuli, presentation order, stimuli classes, ISI, task duration, and display for the IRQ were identical to ST. The question 'Do you have the number (X)' was replaced with 'Is this number greater than 10?' Subjects were instructed to judge whether the number presented in the target area was greater than 10 or not (magnitude judgment) and press the corresponding 'YES' or 'NO' buttons on the fORP to respond. Prior to initiating the IRQ subjects were given two practice trials administered verbally over the scanner intercom to ensure the subject understood the instructions.

Orienting task (OR)

Stimuli, presentation order, stimuli class, ISI, task duration, and display for the OR were also identical to the ST. However, the question 'Do you have the number (X)?' was removed without replacement. Subjects were instructed to attend to each stimulus presented in the target area during the OR task.

After the modification of the ST query in the IRQ task and the outright removal of the ST query in the OR task, there was no longer an act of deception taking place in these two tasks. Thus, the 'Lie' and 'Truth' conditions of the ST were renamed to 'Target' and 'Distracter' conditions in the IRQ and OR tasks.

Image acquisition

MRI scanning was performed on a 3-Tesla Siemens Trio scanner (Iselin, NJ). Functional data were collected with a Blood Oxygenation Level Dependent (BOLD) sequence (TR/TE = 3000/30 ms, FOV = 240 mm, matrix = 64 × 64, slice-thickness/gap = 3/0 mm). For anatomical reference, registration of functional data, and for normalization of functional data to a standard T1 template (Montreal Neurological Institute, MNI) a T1 magnetization prepared, rapid-acquisition gradient echo (MPRAGE, TR/TE = 1630/3.87 ms, FO = 250 mm, matrix = 192 × 256, slice-thickness/gap = 1/0 mm) sequence was used to collect a high-resolution image of each subject's brain. Task stimuli were presented via a video projector (Powerlite 7300; Epson America, Long Beach, CA) and refracted to the subject's visual field with a head-coil mounted mirror.

fMRI data preprocessing

Preprocessing: FMRI and MRI data were preprocessed and analyzed using FMRIB's Software Library (FSL) fMRI Expert Analysis Tool (FEAT) (Smith et al., 2004). Functional data were brain-extracted (Smith, 2002), motion-corrected to the median functional image using b-spline interpolation (4 df), high-pass filtered (60 s), and spatially smoothed (9 mm full width at half maximum (FWHM), isotropic). The anatomical volume was brain-extracted and registered to the standard space T1 MNI template using tri-linear interpolation with FMRIB's Linear Image Registration Tool (FLIRT, 12 df; Jenkinson & Smith, 2001). The median functional image was registered to the anatomical volume, and then transformed to the MNI template.

Statistical analysis of imaging data

Statistical images were created using FEAT with an improved General Linear Model (GLM). Regressors were created by convolving concatenated stimuli time-courses for each number 1 through 9 with the canonical Hemodynamic Response Function (HRF, double gamma). The nine regressors along with their temporal derivatives, and an intercept form were entered into single-subject GLMs for analysis of each task. A contrast of beta-coefficients for the Lie regressor versus the average of the Truth regressors (and Target versus the average of Distracter [IRQ and OR]) were made for each task and resulting images were converted to percent signal change as well as z-statistic maps. The Control-Truth condition, a standard element of a conventional Stim Test, was not included in the analysis. It was not considered comparable to the Truth condition because subjects were instructed that the numbers 1, 2, and 9 could not be used as Lie items.

CIT meta-analysis

Raw data volumes for Lie and Truth conditions from the previously reported GKT1 (Langleben et al., 2002) and the Lie and Repeat-distracter in GKT2 (Langleben et al., 2005) experiments were subjected to the preprocessing steps described in the 'fMRT data preprocessing' section. Contrasts of parameter estimates for Lie and Truth (in GKT1 and GKT2) conditions were entered into a group GLM where a one-way t-test was performed to identify significant activation differences between conditions. The resultant t-map was thresholded at a voxel-height probability of $p < .001$ and cluster-probability of $p < .05$. The thresholded and cluster-corrected volume was then converted to a binary image (0 = non-significant, 1 = significant voxels) for use as a masking volume.

ROI analysis of the ST, IRQ, OR data

ST, IRQ, and OR contrasts were masked by the binary functional result of the CIT Meta-analysis (both described above). Difference in mean percent signal change was calculated from the Lie > Truth and Target > Distracter contrasts within each of the masked ROIs. Scaling factors for percent signal change were comparable between tasks and subjects as the task-design was sparse event-related, allowing all events to be isolated, all events were of the same duration, and the design matrix was identical between each task and each subject.

RESULTS

Behavioral data

The error rate for all three subjects in all tasks was 0%.

CIT meta-analysis (Figure 1)

The Lie > Truth ($N = 40$) comparison revealed significant activations in bilateral inferior frontal gyrus (IFG), bilateral inferior parietal lobe (IPL, primarily supramarginal gyrus [SMG]), a cluster extending between the superior frontal gyrus (SFG) and the anterior cingulate (ACC), the dorsal region of the ACC, bilateral middle temporal gyrus (MTG), and the precuneus (BA 7) (Table 1). All reported coordinates are in the MNI standard space. The resultant meta-analysis volume was thresholded at a voxel-height of $p < .001$, and a cluster-probability of $p < .05$ was used to control for Type I and II errors.

Subject 1 (Figure 2A)

During the ST, Subject 1 exhibited increased activation in the Lie > Truth comparison in the bilateral IFG, IPL/SMG, SFG, dorsal ACC, and the right MTG. During the IRQ, increased Target-activation was present in left IFG. Lastly, during the OR increased Target-activation was present in the bilateral IFG, IPL/SMG, SFG, dorsal ACC, and the left MTG. There was an overlap between ST (Lie) and OR (Target) related activation in six regions. The overlap between ST (Lie) and IRQ (Target) was present only in the left IFG.

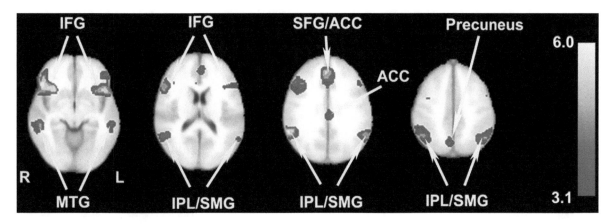

Figure 1. fMRI results of CIT Meta-Analysis ($N = 40$) Lie > Truth Contrast. Results are z-statistic maps thresholded at voxel-height probability of $p < .001$, and cluster-probability of $p < .05$, displayed over the MNI T1 anatomical template in radiological convention (the right side of the brain is on the viewer's left). Significant clusters of activation are located in bilateral IFG, IPL, SFG, Dorsal ACC, MTG, and precuneus.

TABLE 1
fMRI results of the CIT Meta-analysis ($N = 40$)
Lie > Truth contrast

Region	Brodmann area	X	Y	Z	Z_{max}
IFG L	44, 45, 47	−44	20	−4	5.38
IFG R	44, 45, 47	50	26	−4	5.95
IPL/SMG L	40	−58	−52	32	5.17
IPL/SMG R	40	60	−48	30	5.46
ACC	8, 32	2	32	36	5.2
ACC (dorsal)	24, 31	−4	−26	28	4.11
MTG L	20, 21, 39	−58	−32	−12	3.92
MTG R	20, 21, 39	52	−34	−8	4.48
Precuneus	7	8	−68	42	4.05

Coordinates (X, Y, Z) are in MNI standard space. Results are thresholded at voxel-height probability of $p < .001$ and a cluster-probability of $p < .05$.

Subject 2 (Figure 2B)

During the ST, Subject 2 exhibited increased activation in the Lie > Truth comparison in the bilateral IFG, IPL/SMG, SFG, dorsal ACC, and MTG. During the IRQ, increased Target-activation was primarily in the left IFG, left MTG, and precuneus. Lastly, during the OR, increased Target-activation was present in every ROI: bilateral IFG, bilateral IPL/SMG, SFG, dorsal ACC, bilateral MTG, and precuneus. The overlap between ST (Lie) and OR (Target) related activation was in all ROIs except the precuneus. Increased Lie/Target activation was shared between ST and OR only in the left IFG and left MTG.

Subject 3 (Figure 2C)

In this subject, increased activation in the Lie > Truth (or Target > Distracter) comparison was present in all ROIs in all three tasks: bilateral IFG, IPL/SMG, SFG, dorsal ACC, MTG, and precuneus.

DISCUSSION

Meta-analysis of the Lie vs. Truth contrast in the two CIT paradigms produced a prefronto-parietal pattern that included the bilateral IFG and IPL, bilateral MTG, as well as the SFG and ACC. A recognized flaw of the CIT is the stimulus familiarity confound i.e., the concealed Lie stimulus was more familiar to the participants than the Truth stimuli (Langleben et al., 2002). Nevertheless, the fact that similar prefronto-parietal activation pattern has also been reported in fMRI studies that have used non-CIT deception paradigms (Abe, Suzuki, Mori, Itoh, & Fujii, 2007; Kozel et al., 2005; Nunez, Casey, Egner, Hare, & Hirsch, 2005; Spence et al., 2001) supports our use of these regions in subsequent *a priori* ROI analyses. The ST reported here is a CIT that has been controlled for familiarity, yet it shows an activation pattern similar to the CIT meta-analysis across all three subjects, downplaying the importance of the familiarity confound.

The OR task is essentially a passive viewing of the ST; there is no query and no behavioral

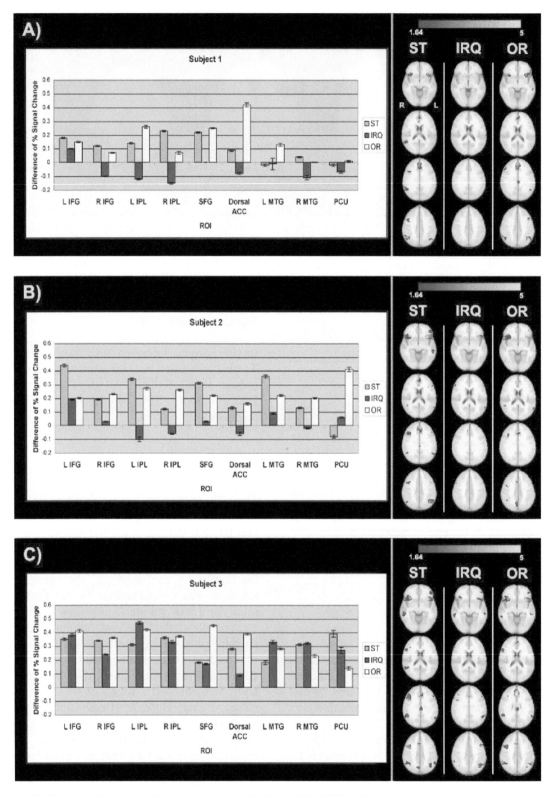

Figure 2. ST, IRQ, and OR. Subject 1 (A), Subject 2 (B), and Subject 3 (C): (LEFT) Difference in mean percent signal change between Lie > Truth conditions (ST) and Target > Distracter conditions (IRQ and OR). (RIGHT) fMRI whole-brain results of Lie > Truth (ST) and Target > Distracter (IRQ and OR) contrasts thresholded at a voxel-height probability of $p < 0.05$, uncorrected for multiple comparisons and masked by thresholded CIT meta-analysis Lie > Truth contrast (Figure 1). Results are z-statistic maps displayed over the MNI T1 anatomical template in radiological convention (Subjects' right is on viewer's left).

response is required. These manipulations convert the ST Lie and Truth items into the Target and Distracter stimuli of the OR. The Target vs. Distracter contrast in the OR task shows prefronto-parietal, ACC and MTG activation, all virtually identical to the ST Lie vs. Truth pattern. These findings may call for a departure from the prevailing hypothesis that postulates that prefrontal activation is specific to deception. One caveat is that this similarity could have been caused by an order effect: since ST always preceded the OR, the cognitive set of lying established by the ST may have been maintained despite the change in instructions from deception to passive observation. However, the fact that for two out of three subjects the ST and OR pattern similarities did not persist in the IRQ, which also followed the ST, argues against this interpretation. In addition, even this more conservative interpretation demonstrates that pairing of a motor response to item recognition is unnecessary in eliciting the pattern observed in the ST. Confirmation of the similarities between the fMRI response pattern during a lie and a generally salient item in a counterbalanced experiment would suggest that the CIT is indeed a test of some of the cognitive operations involved in deception, such as orienting, endogenous attention, and matching to target. Such confirmation would also suggest that a CIT is unable to specifically indicate a deceptive response.

Unlike the results of the ST and the OR, the IRQ results varied across subjects. The expected prefronto-parietal activation was observed in Subject 3 only. In Subjects 1 and 2, shared activation during irrelevant responding to the target item was only found in the left IFG. The order effects between the ST and the IRQ are the same as between the ST and the OR. Since the OR and IRQ were counterbalanced, the difference in fMRI response to the target item between OR and IRQ could not be attributed to order effects. We hypothesize that the cognitive interference of the irrelevant response (magnitude judgment) to the formerly concealed item could effectively disrupt the cognitive set/order effect imposed by the ST. The applied significance of this interpretation is that cognitive interference may be an effective countermeasure to CIT-based fMRI interrogation. The direct comparison made between target and distracter in analysis of IRQ should, in theory, yield no difference between conditions, as all trials involve presumably the same cognitive and behavioral response pattern. We hypothesize that the remarkable persistence of the left IFG during

the IRQ for both subjects represents maintenance of target stimulus-saliency throughout the testing session. The left IFG has been shown to be involved with retrieval of semantic knowledge in a variety of domains and is sensitive to increased working memory load (Badre, Poldrack, Pare-Blagoev, Insler, & Wagner, 2005; Brass & von Cramon, 2004; Feredoes, Tononi, & Postle, 2006; Jonides & Nee, 2006; Lauro, Tettamanti, Cappa, & Papagno, 2007; Thompson-Schill, D'Esposito, Aguirre, & Farah, 1997). Despite the distractive effect of the IRQ in inhibiting the full deceptive cognitive response, activation of the left IFG due to stimulus-saliency, and consequential working memory competition to break behavioral set, may be predictive of a concealed target. Persistence of the left IFG in the IRQ task is in agreement with our previous report of increased left IFG response during Lie conditions (Langleben et al., 2005).

The data presented here are derived from only three subjects and is therefore preliminary. Their main value is in reporting a possible method for formal study of a traditional deception model through the use of a systems neuroscience approach. Replication of these findings in a larger sample would provide imaging data amenable to a random effects analysis as well as sufficient reaction-time data. The manipulations of the CIT we reported have several limitations which could be addressed by future larger studies: first, simultaneous removal of the query and response requirement in the OR precludes it from dissociating the contributions of the response and of the query to the CIT fMRI pattern. The potential of order effects and habituation could be avoided by counterbalancing across all three tasks rather than only the IRQ and OR. Between-tasks comparisons could not be made due to confounds of novelty and repetition of stimuli, a limitation that could also be overcome with a larger sample size. Finally, the experiments proposed here might not answer the fundamental question of whether there is a brain response pattern characteristic of deception that is independent of the structure of the task used to elicit it.

In conclusion, the meta-analysis data presented here confirm the relevance of prefronto-parietal brain activation to deception. Together with the results of the ST, these preliminary data support the internal validity of the fMRI-adapted CIT model of deception. Deceptive behavior is not essential for the CIT-type response pattern, while cognitive distraction can significantly attenuate it. The manipulations presented here are a blueprint for the use of cognitive neuroscience methodology to decode the contributions of basic cognitive operations to the brain pattern of deception, and to help the search for the existence of an activation pattern specific or even unique to deception. In the cases reported here, the left IFG activation has been resistant to cognitive distraction, suggesting that it may be a marker of deception and/or concealed information that is independent of the structure of the query used to elicit deceptive behavior. No model to-date has convincingly dissociated a brain fMRI signal of deception from the brain activity associated with basic processing of the cognitive task used to elicit deceptive behavior. Such a model may require consideration of social interaction, exercise of agency, and moral judgment during deceptive behavior (Frith, 2007; Greene, Nystrom, Engell, Darley, & Cohen, 2004; Watson, 2001).

REFERENCES

Abe, N., Suzuki, M., Mori, E., Itoh, M., & Fujii, T. (2007). Deceiving others: Distinct neural responses of the prefrontal cortex and amygdala in simple fabrication and deception with social interactions. *Journal of Cognitive Neuroscience, 19*(2), 287–295.

Aguirre, G. K., & D'Esposito, M. (1999). Experimental design for brain fMRI. In C. T. W. Moonen, & Bandettini, P.A. (Eds), *Functional MRI* (pp. 369–380). New York: Springer.

Badre, D., Poldrack, R. A., Pare-Blagoev, E. J., Insler, R. Z., & Wagner, A. D. (2005). Dissociable controlled retrieval and generalized selection mechanisms in ventrolateral prefrontal cortex. *Neuron, 47*(6), 907–918.

Brass, M., & von Cramon, D. Y. (2004). Selection for cognitive control: A functional magnetic resonance imaging study on the selection of task-relevant information. *Journal of Neuroscience, 24*(40), 8847–8852.

Dale, A. M. (1999). Optimal experimental design for event-related fMRI. *Human Brain Mapping, 8*(2–3), 109–114.

Elaad, E., & Kleiner, M. (1986). The stimulation test in polygraph field examinations: A case study. *Journal of Police Science & Administration, 14*(4), 328–333.

Feredoes, E., Tononi, G., & Postle, B. R. (2006). Direct evidence for a prefrontal contribution to the control of proactive interference in verbal working memory. *Proceedings of the National Academy of Sciences of the United States of America, 103*(51), 19530–19534.

Frith, C. D. (2007). The social brain? *Philosophical Transactions of the Royal Society of London. Series B, Biological Sciences, 362*(1480), 671–678.

Furedy, J. J., Gigliotti, F., & Ben-Shakhar, G. (1994). Electrodermal differentiation of deception: The effect of choice versus no choice of deceptive items. *International Journal of Psychophysiology, 18*(1), 13–22.

Greene, J. D., Nystrom, L. E., Engell, A. D., Darley, J. M., & Cohen, J. D. (2004). The neural bases of cognitive conflict and control in moral judgment. *Neuron, 44*(2), 389–400.

Jenkinson, M., & Smith, S. (2001). A global optimisation method for robust affine registration of brain images. *Medical Image Analysis, 5*(2), 143–156.

Jonides, J., & Nee, D. E. (2006). Brain mechanisms of proactive interference in working memory. *Neuroscience, 139*(1), 181–193.

Kozel, F. A., Johnson, K. A., Mu, Q., Grenesko, E. L., Laken, S. J., & George, M. S. (2005). Detecting deception using functional magnetic resonance imaging. *Biological Psychiatry, 58*(8), 605–613.

Langleben, D. D., Schroeder, L., Maldjian, J. A., Gur, R. C., McDonald, S., Ragland, J. D., O'Brien, C. P., & Childress, A. R. (2002). Brain activity during simulated deception: An event-related functional magnetic resonance study. *Neuroimage, 15*(3), 727–732.

Langleben, D. D., Loughead, J. W., Bilker, W. B., Ruparel, K., Childress, A. R., Busch, S. I., & Gur, R. C. (2005). Telling truth from lie in individual subjects with fast event-related fMRI. *Human Brain Mapping, 26*(4), 262–272.

Lauro, L. J., Tettamanti, M., Cappa, S. F., & Papagno, C. (2007). Idiom comprehension: A prefrontal task? *Cerebral Cortex, 18*(1), 162–170.

Lykken, D. T. (1991). Why (some) Americans believe in the lie detector while others believe in the guilty knowledge test. *Integrative Physiological and Behavioral Science, 26*(3), 214–222.

Matte, J. A. (1996). *Forensic Psychophysiology Using the Polygraph: Scientific Truth Verification – Lie Detection* (1st ed. Vol. 1). Williamsville, NY: JAM Publications.

Nunez, J. M., Casey, B. J., Egner, T., Hare, T., & Hirsch, J. (2005). Intentional false responding shares neural substrates with response conflict and cognitive control. *Neuroimage, 25*(1), 267–277.

Smith, S. M. (2002). Fast robust automated brain extraction. *Human Brain Mapping, 17*(3), 143–155.

Smith, S. M., Jenkinson, M., Woolrich, M. W., Beckmann, C. F., Behrens, T. E., Johansen-Berg, H., Bannister, P. R., De Luca, M., Drobnjak, I., Flitney, D. E., Niazy, R. K., Saunders, J., Vickers, J., Zhang, Y., De Stefano, N., Brady, J. M., & Matthews, P. M. (2004). Advances in functional and structural MR image analysis and implementation as FSL. *Neuroimage, 23*(Suppl. 1), S208–219.

Spence, S. A., Farrow, T. F., Herford, A. E., Wilkinson, I. D., Zheng, Y., & Woodruff, P. W. (2001). Behavioural and functional anatomical correlates of deception in humans. *Neuroreport, 12*(13), 2849–2853.

Spence, S. A., Hunter, M. D., Farrow, T. F., Green, R. D., Leung, D. H., Hughes, C. J., & Ganesan, V. (2004). A cognitive neurobiological account of deception: Evidence from functional neuroimaging. *Philosophical Transactions of the Royal Society of London. Series B, Biological Sciences, 359*(1451), 1755–1762.

Thompson-Schill, S. L., D'Esposito, M., Aguirre, G. K., & Farah, M. J. (1997). Role of left inferior prefrontal cortex in retrieval of semantic knowledge: A reevaluation. *Proceedings of the National Academy of Sciences of the United States of America, 94*(26), 14792–14797.

Verschuere, B., Crombez, G., De Clercq, A., & Koster, E. H. (2004). Autonomic and behavioral responding to concealed information: Differentiating orienting and defensive responses. *Psychophysiology, 41*(3), 461–466.

Watson, G. (2001). Free agency. In L. W. Ekstrom (Ed.), *Agency and responsibility: Essays on the metaphysics of freedom* (pp. 72–106). Boulder, CO: Oxford Press.

NEUROCASE
2008, 14 (1), 68–81

Looking for truth and finding lies: The prospects for a nascent neuroimaging of deception

Sean A. Spence and Catherine J. Kaylor-Hughes

Academic Clinical Psychiatry, University of Sheffield, Sheffield, UK

Lying is ubiquitous and has acquired many names. In 'natural experiments', both pathological lying and truthfulness implicate prefrontal cortices. Recently, the advent of functional neuroimaging has allowed investigators to study deception in the non-pathological state. Prefrontal cortices are again implicated, although the regions identified vary across experiments. Forensic application of such technology (to the detection of deceit) requires the solution of tractable technical problems. Whether we 'should' detect deception remains an ethical problem: one for societies to resolve. However, such a procedure would only appear to be ethical when subjects volunteer to participate, as might occur during the investigation of alleged miscarriages of justice. We demonstrate how this might be approached.

Keywords: Deception; Pathological lying; Prefrontal cortex; Psychiatric Neuroimaging.

INTRODUCTION

In an imagined future, where incontrovertible DNA evidence places a single subject at the scene of a crime, and irrefutable closed-circuit television (CCTV) footage records, plays and replays, the movements his body made, would there be anything left for investigators to explain? The conclusion might appear obvious: the person who was present at the time of the crime is 'guilty'. We reply: 'not necessarily'. Despite the enormous advances in, and application of, DNA 'finger-printing', and our currently expanding surveillance culture, there is perhaps still some nagging doubt which will arise in those of us concerned with states of mind and consciousness. For, while a body might undoubtedly leave evidence that 'it' was present at the scene of a crime, and while film might recapitulate the movements which that body 'made', we are not simply logging the motions of machines here. In the central case of human conduct (especially when that conduct is abnormal and/or immoral) we are crucially concerned with intentions, states of mind, with what the subject was thinking. Was he acting as an agent when he lifted his hand? Did he 'intend' to do what he did, when he did it?

Admittedly, this may be largely a matter of interest to those concerned with the mind/brain problem. The evidence of forensic science can be most compelling and the observation of someone 'caught in the act' understandably suggests that there is no defence. However, we are cognisant that even this 'perfect' scenario is insufficient to condemn someone unreservedly. What if another person had compelled the act? (*'I robbed the bank because they were holding my wife hostage'.*) What if actions were based upon misunderstandings? (*'He reached for the hammer; I thought he was going to hit me'.*) What if the subject had been

We are grateful to all our colleagues in SCANLab at the University of Sheffield for their contributions to discussions of these data and we thank Mrs Jean Woodhead for her kind assistance in manuscript preparation.

Address correspondence to Professor Sean A. Spence, Academic Clinical Psychiatry, University of Sheffield, The Longley Centre, Norwood Grange Drive, Sheffield S5 7JT, UK (E-mail: S.A.Spence@Sheffield.ac.uk).

DOI: 10.1080/13554790801992776

deluded at the time? *('When I looked into his eyes I knew he was the Devil.')* What if his consciousness were disturbed? *('There was a strange rotting smell and then my mind went blank'.)* And what if the evidence against him has been falsified?

We posit that we are likely to remain interested in the mental states of those who are accused of antisocial acts for some time to come; that contemporary and near-term neuroscience may provide tentative insights into such situations. However, in this paper, we begin at the end. We consider what might happen *after* such events have occurred. We take as our example the brain imaging studies that have probed deception; we discuss their findings, caveats and likely future applications. Can we tell whether someone is reporting his actions truthfully? First, we consider the nature of deceit.

TELLING LIES

Lie: a false statement made with the intention of deceiving. (Chambers Concise Dictionary, 1991)

Despite injunctions to the contrary, humans engage in lies and deceit on a regular basis. While children may be told that lying is bad, they might also be told not to be rude, not to hurt others' feelings or to appear insensitive. In later life, a totally frank exchange of views can leave both parties feeling bruised and it is commonly conceded that a certain amount of 'tact and diplomacy' or 'impression management' is necessary in navigating one's way among con-specifics (see Vrij, 2000, for an accessible overview).

[O]ne must know best how to colour one's actions and to be a great liar and deceiver. (Niccol Machiavelli, *The Prince*, 1999, p. 57)

In social psychological research, two findings are notable: first, lies are common, particularly those told to partners and parents and, second, humans are poor at detecting when they are being deceived (Vrij, 2000). While many subjects might believe that they are astute detectors of deception, the empirical data point to most of us operating at, or a little above, the level of chance and this finding is replicated across groups of professionals (including police, judges, doctors and lawyers; Ekman & O'Sullivan, 1991). While some secret service personnel and some offenders may detect deceit at levels 'above chance' (Ekman &

O'Sullivan, 1991; Granhag & Stromwall, 2004), most humans share a 'truth bias', i.e., we tend to assume that others are speaking truthfully. There may be sound (admittedly, teleological) reasons for this: not to believe others might render one paranoid, all information would require independent verification, and life would be lived most inefficiently. As demonstrated by contemporary economic scares, and political disengagement, a lack of trust greatly impedes social intercourse; it is mutually disadvantageous (Galbraith, 2005; Oborne, 2006). Also, a breakdown of trust can make other aspects of life and health falter: in recent studies, the magnitude of economic polarisation (inequality) has been implicated in increasing mistrust, depression and violence in some Western democracies (Layard, 2005; Wilkinson, 2005). Thus, trusting others, though it might seem 'naïve', appears to constitute a healthier way of life.

Nevertheless, in the struggle for survival, even advantageous evolutionary processes may precipitate vulnerabilities in the 'wrong' environment, and it has been posited (by game theorists) that our truth bias leaves open the way for a small number of 'free-loaders', 'cheats' or psychopaths to prosper (Mealey, 1995). The theory is that while most subjects adopt a trusting policy a small number of cheats might 'get away' with cheating, so long as they are diffusely distributed in the population. However, if a subject becomes recognized as a cheat then their survival depends upon either a change of location or a radical change of policy! Notice, also, that a society largely composed of cheats and misanthropes would likely stall (who would enter into any kind of binding agreement with others?).

Furthermore, the attempt to deceive another tells us something interesting about the putative deceiver. It suggests that he understands, whether implicitly or explicitly, that different people can hold different perspectives on the same subject, that agents experience different thoughts (O'Connell, 1998). Hence, the deceiver possesses a 'theory of mind'; this is a prerequisite for his belief that what he says or does (his deceit) is not immediately apparent to his interlocutor; for his belief that he might create a false belief in the mind of another. The developmental literature suggests that children develop the ability to attempt deceit at about 3–4 years of age (O'Connell, 1998). This is congruent with their developing theory of mind abilities, i.e., they come to understand that they can know something that their carer does not: there is the possibility of managing information.

Hence, to engage in deceit, a deceiver has to understand that different agents can believe different things; they have to understand (at some level) what philosophers call 'intentionality':

> I know p.
> He does not know p.
> I know that he does not know p.
> I shall make him think q, etc.

Hence, attempted deceit also tells us something about the moral state of the deceiver; it informs our understanding of his responsibility (below).

THE EVOLUTION OF DECEPTION

While the use of camouflage and distraction has a very long evolutionary past, reaching back towards relatively primitive organisms (Giannetti, 2000), some 'knowing' use of deception (with a degree of insight into the deceived other's perspective) appears to reach its apogee among primates. The work of Byrne and colleagues is very informative here (e.g., Byrne, 2003). When human researchers observe non-human primate colonies, the subjects of such observation can sometimes be seen to seemingly 'deliberately' mislead their con-specifics. It seems as if some primates can deploy normal elements of their behavioural repertoires in such a way as to mislead their opposition (to their own advantage). Such behaviour has been termed 'tactical deception' and, allowing for the frequency with which human researchers will have studied different primate groups, it appears that the frequency of tactical deception increases through the evolutionarily more advanced of the primates (i.e., gorillas, bonobos and chimpanzees), bearing a statistical relationship to the volume of their neocortices. Hence, the implication is that 'higher' centres of the primate brain facilitate this recently emergent behaviour: the 'knowing' use of deception (Byrne & Corp, 2004). Of course, without recourse to other primates' mental states, we cannot know to what extent their actions are indeed conscious or premeditated, but certain of them appear 'knowing'.

Another implication of such work is that it places human behaviour at the end of an evolutionary arc; hence, human deception, in all its many guises (Table 1), may be similarly supported by adaptive

TABLE 1
Varieties of 'deceiving' (from *Roget's Thesaurus*, 2004, p. 220)

Related terms
Deceive, delude
Beguile, sugar the pill, gild the pill, give a false impression, belie
Let down, disappoint
Pull the wool over one's eyes, blindfold, blind
Kid, bluff, bamboozle, hoodwink, hoax
Throw dust in the eyes, create a smoke screen, lead up the garden path, mislead
Spoof, mystify, misteach
Play false, leave in the lurch, betray, two-time, double-cross, be dishonest
Steal a march on, be early,
Pull a fast one, take one for a ride, outsmart, be cunning
Trick, dupe
Cheat, cozen, con, swindle, sell, rook, do, defraud
Diddle, do out of, bilk, fleece, rip off, shaft, shortchange, obtain money by false pretences
Juggle, conjure, palm off, foist off, fob, fob off with, fiddle, wangle, fix
Load the dice, mark the cards
Counterfeit, fake

N.B. This represents only a fraction of the total terms related to 'deception ..trickery ..sleights of hand ..traps ..shams, etc.'.

changes in cognitive neurobiological architecture. If we hypothesise that this is the case, then we may extend our theory to suggest that 'higher' brain systems will be engaged when humans tell lies (Spence, 2004).

THE PSYCHIATRY OF DECEPTION

It seems reasonable to hypothesize that trust and truth bias are most exposed in those encounters where the phenomena to be discussed are subjective; where only one party has direct access to the information concerned and when that material comprises the contents of their own state of consciousness. Hence, a discipline such as psychiatry, concerned as it is with abnormalities of phenomenology (the subjective state), must place especial emphasis upon the relationship between speakers and their veracity. Listening very closely to what others say is a central component of the medical process but it can misfire. Differences in culture, class, education, and assumptions about the world may all lead to honest confusions: misunderstandings. However, there will also be occasions when one or other party sets out deliberately to mislead their interlocutor, or falls into such a procedure as a mode of impression management. Doctors have traditionally been accused of failing to disclose a patient's diagnosis, especially when the prognosis is poor; some have administered placebos, purportedly benignly, but certainly paternalistically, misleading their subjects (Sokol, 2006). Patients may conceal their non-adherence to treatments, their disagreement with the doctor or their seeking of treatment elsewhere, but sometimes deception is more integral to their state (Table 2, left column; Hughes, Farrow, Hopwood, Pratt, Hunter, & Spence, 2005).

We can identify two categories of psychiatric syndrome associated with deception: those where deception is central to the diagnosis itself (as in antisocial and psychopathic personalities) and those where deception arises as a consequence of something else, contingent upon another disturbance of functioning (e.g., in substance misuse or eating disorders, where patients may be embarrassed or ashamed of their thoughts and actions and therefore seek to conceal them). To acknowledge the occurrence of deception is not necessarily to make moral judgements of patients, it is to acknowledge that human communication can be complex and multi-layered and that, though we do not wish to judge others, we may have to anticipate their intentions. In some cases the deception may harm the patient herself (e.g., suicidal intent is denied but then acted upon), sometimes the harm is directed toward others (e.g., a sexual victim is being 'groomed' while intent is denied). There are many possible scenarios within the psychological

and psychiatric realm (see Table 2; and Hughes et al., 2005). It also follows that if doctors are little better than chance in their detection of deception (Ekman & O'Sullivan, 1991), then there may be serious consequences for either their patients or others when they misattribute truthfulness.

PATHOLOGICAL LYING

If lying is a 'normal' aspect of human behaviour, can it ever be termed 'pathological'? The literature describes two broad possibilities:

1. Those cases where lying is used to harm *others*, to con and to cheat them (e.g., Yang et al., 2005); and
2. Those cases where 'stories' are told, which do not seem designed to gain tangible reward. Such liars seem to harm *themselves* by their habits; their motivations remain obscure (e.g., Munchausen's syndrome, pseudologia fantastica; Bass, 2001).

There have been relatively few empirical studies of the biology of pathological lying but one sequence of work has been informative. Raine's group have studied antisocial people living in the community in Los Angeles and have found that, compared with control groups, they exhibit reduced prefrontal grey matter. However, among those antisocial people who exhibit a pattern of repeated lying and cheating (for instance, through fraud) they have found an excess of prefrontal white matter (Yang et al., 2005), especially in orbitofrontal cortices (Yang et al., 2007), even relative to other antisocials. Given what we have alluded to (above) with respect to primate evolution, and the sociological, game theory notion that psychopathy may represent a persistent, stable evolutionary strategy, if minimally represented in human societies (Mealey, 1995), it is conceivable, as suggested by Raine's group, that antisocial pathological liars are somehow 'helped' in their deception by a facility bestowed upon them by differences in prefrontal white matter structure. This would be congruent with an older literature suggesting that such liars exhibit enhanced verbal skills (Ford, King, & Hollender, 1988). This is a fascinating line of work and argument. However, we must acknowledge one caveat: we cannot say at present whether such prefrontal findings represent 'cause' or 'effect' (Spence, 2005). Does brain structure predispose to lying or does repeated lying alter brain structure? It is too soon to say.

TABLE 2
Psychiatric syndromes invoking deception, directly or indirectly

Deception central to diagnosis	Deception a possible consequence
Conduct disorder	Eating disorders
Antisocial personality disorder	Alcohol addiction
Psychopathy	Substance misuse disorders
Munchausen's syndrome	Paraphilias
Munchausen's syndrome by proxy	Obsessive compulsive disorder
Malingering	
'Pathological lying'	
Instrumental psychosis	

N.B. There are other psychiatric concepts where the question of deception is either controversial, ambiguous or the 'victim' is thought to be the subject herself, e.g., conversion and dissociation syndromes (hysteria), the psychodynamic 'defence' mechanisms, 'self-deception', and delusions.

PATHOLOGICAL TRUTHFULNESS

While some psychiatric disorders are associated with a propensity towards deception others may be associated with a 'failure' to deceive. It has long been recognised that children with autism exhibit difficulties in deceiving others, which may be related to deficits in their abilities to 'mind read' or appreciate others' perspectives (a deficit in their 'theory of mind'; e.g., Happe, 1994). While it is likely that the failure of cognitive processing revealed by such pathological truthfulness implicates frontal lobe processes (Fletcher et al., 1995), autism is of course associated with other psychopathological disturbances and impairments of development. Hence, it would be interesting to ask whether pathological truthfulness can ever emerge among adults who have previously experienced 'normal' development. Two lines of enquiry are informative:

1. In some patients with orbitofrontal (OFC) lesions a behaviour termed 'pseudo-psychopathy' has been described, characterised by 'outspokenness', lack of 'tact and restraint', and being 'brash and disrespectful', 'open and frank' (Blumer & Benson, 1975).
2. There is recent empirical evidence that such patients (with OFC lesions) are unnecessarily confiding, overly intimate in the details which they share with others (Beer, Heerey, Keltner, Scabini, & Knight, 2003).

Taken together, these strands of evidence (admittedly emerging from rather small literatures) point to a relationship between prefrontal, especially orbitofrontal, lobe function and a subject's veracity. While those who lie to harm others exhibit increased OFC white matter (Yang et al., 2007), those who experience lesions (as adults) in similar regions are rendered 'tactless' and inappropriately truthful (Blumer & Benson, 1975). On the background of what we have learned from non-human primate studies it seems permissible to hypothesise that OFC regions may be implicated in the human capacity for bearing true or false witness.

NEUROIMAGING OF DECEIT

Much of the evidence that we have considered thus far relied upon 'natural experiments' for its acquisition: the observations of field researchers among non-human primates, the consequences of brain damage to formerly 'normal' humans, the unknown aetiology of those difficulties in communication encountered by people with autism. To rely upon such sources of information in constructing a biological science of deception would condemn researchers to opportunistic acquisition of 'special' cases; its advance would be largely stochastic. However, the recent advent, and widespread availability, of functional neuroimaging technology has allowed us the opportunity of studying deception in the healthy living brain in a planned, step-wise progression. With modern brain imaging techniques we really can begin to ask: *which brain systems are implicated in telling a lie?*

The Sheffield model

From the beginning of our work in this area (ca. 2000) we have followed a rather simple model, hypothesizing that deception is reliant upon the prefrontal cognitive executive for its execution. To be more specific, we have posited that key regions of prefrontal cortex will be implicated in different aspects of the deceptive process:

1. We posited, on the basis of a wide range of basic neurological and functional neuroimaging data (demonstrating the role of dorsolateral prefrontal cortex (DLPFC) in the generation of novel responses or responses where the subject is granted latitude about what to choose; e.g., Spence, Hunter, & Harper, 2002), that DLPFC would be implicated when liars elaborate new 'outright' lies, i.e., when they 'tell stories'.
2. However, we also posited that there was another process that was more basic to the execution of deception: the ability to withhold the truth (concealment). To consider this phenomenon we posited that: where a subject knows the answer to a question, that answer forms a relatively pre-potent response to that question. So, in telling a lie, the liar must crucially withhold the pre-potent response, the truth. We posited that such processes would be most exposed under those conditions where the liar is constrained to answering either 'yes' or 'no', and where there is an emphasis upon speed of response. Hence, we posited that inhibitory prefrontal regions should be implicated in this form of lying, and candidate regions would be orbitofrontal cortex (OFC), more specifically

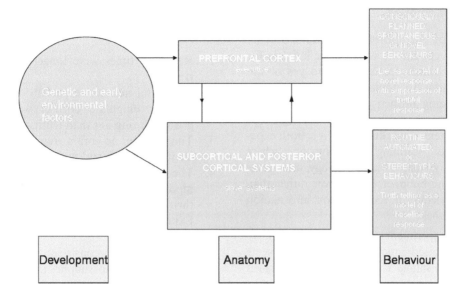

Figure 1. A cartoon showing the Sheffield model, first outlined in Spence (2004). The influence of developmental factors (left) impacts upon present state cognitive neurology (centre) and the ability of executive centres (middle, top) to exert control over 'lower' centres. Emergent behaviours are shown (right), where truthful responding resembles a pre-potent, automatic response (below) and a lie an elaboration of new material (above).

the ventrolateral prefrontal cortex (VLPFC; an area known to be involved in response alternation and suppression; Butters, Butter, Rosen, & Stein, 1973; Iversen & Mishkin, 1970; Starkstein & Robinson, 1997). As we have already noted, lesions of OFC may render adults pathologically truthful.

3. Finally, we incorporated a finding that has long been recognized in the polygraph and social psychology literatures: that lies tend to incur longer response times than truths (Vrij, 2000). Such a finding is consistent with our model, in that we posit that 'higher' elements of the cognitive architecture are engaged by deception, their involvement incurring a processing cost – increased response time.

Placing all these elements in the model, we arrived at the framework shown in Figure 1. In it, we move from developmental factors (on the left) to current neuroanatomy (middle) and emergent behaviours (on the right). We should acknowledge a debt to Tim Shallice's (1988) concept of the Supervisory Attention System (prefrontal executive regions modulating 'lower' motor centres, as compressed into the middle section of the framework), and we should admit that the left hand side ('development') is theoretically under-specified, though we should

foresee genetic and early environmental factors contributing here. Indeed, it is between the 'development' and 'anatomy' columns that we would currently place the Raine group's work on pathological liars (exhibiting increased orbitofrontal white matter (above)).

Furthermore, our model predicts certain specific outcomes: lies will be associated with greater prefrontal activation than telling the truth; truthful responses should not be associated with any areas of greater prefrontal activity than deceptive responding; lesions of inhibitory prefrontal regions should 'release' truthful responses; and lesions of prefrontal regions involved in response elaboration (even if transient, as with transcranial magnetic stimulation) should impede new lies but not affect truths.

We should make two further points explicit: (a) the relationship between items in the middle and right hand columns is merely functional and much simplified (e.g., prefrontal cortex would not enact lies 'on its own', it must act 'via' other brain structures), (b) the role of the executive (middle column, top) is both to facilitate and to inhibit, it may bias the lower systems towards new, novel responses (via DLPFC) or suppress their prepotent response tendencies (via VLPFC). We hope the model is justified by the following examples.

Falsifying memories

The kernel of our approach has been to study subjects under conditions where their latitude for response is maximally constrained, to yes or no. We have chosen to focus very much on the information being probed at the expense of the interaction between subjects. This may be critiqued, as a potential weakness, but it is both expedient in the scanning environment, where interactions are mediated via technology, and serves to emphasize the brain of the deceiver rather than that of their 'victim'; it focuses upon information known and withheld.

In our first experiment (Spence et al., 2001), we probed recent episodic memory. We asked subjects which of 36 very simple actions they had performed on the day they were studied. Each question was asked repeatedly in the MR scanner, under a counterbalanced design, and subjects were instructed to give what they regarded as the truthful response in the presence of one cue and what they regarded as a lie in the presence of another (they pressed 'yes' or 'no' buttons on a keyboard). A 'stooge' in the observation room attempted to 'tell' which responses were true or false. One of our team had interviewed subjects prior to the study and obtained a 'ground truth' account of actions performed by each subject on that day (before they knew the relevance of these questions). Hence, we were able to check the consistency of our subjects' responses throughout

the procedure. When we analysed our data we found the following (Figure 2):

1. Response times were longer during lies than truths;
2. Prefrontal activity (especially in VLPFC and anterior cingulate cortex) was significantly greater during lies than truths; and
3. There were no brain regions where truths elicited greater activity than lies;
4. The results were the same when repeated with a modified protocol (questions presented through earphones rather than on a screen; Expt 2, Figure 2).

These findings are consistent with our *a priori* hypotheses; they also support our basic theoretical premise: that truthful responding comprises a relative baseline in human communication and deceit a super-ordinate function, requiring the intervention of cognitive executive systems (thereby incurring a processing cost, manifest as longer response times).

Can this method be applied to forensic questions?

In more recent work we have had the opportunity of applying this simple methodology in rather more serious circumstances: in people who assert that they have been the subjects of miscarriages of justice. In this case, if we are to apply our basic

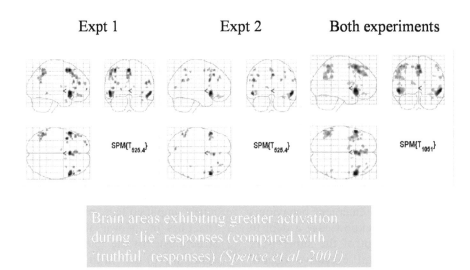

Figure 2. Images showing areas of the brain where lying elicited greater activation than truthful responding in Spence et al. (2001). Responses were made using a keyboard (yes/no) and stimuli were presented visually as questions on a screen (left; Expt 1) or aurally via headphones (middle, Expt 2). A conjunction analysis of both datasets is also shown (right). In each image there is prominent activation of bilateral ventrolateral prefrontal cortices (Brodmann Area 47).

method then each case becomes a single-case study. This is because 'ground truth' relates to a unique event: a crime, with its particular circumstances and time-lines. It follows that apart from basic hypotheses regarding cognitive neurobiological architecture (e.g., the involvement of VLPFC during lie responses) there is no 'control' group for such a study (no one else will have been implicated in precisely the same scenario as the subject whom we study). Now we are dealing with long-term episodic memory but we can adapt our approach to ask questions regarding those specific points upon which there is disagreement (the conflicting views of accusers and accused). Nevertheless, we are also reliant upon several less predictable factors:

1. A clear narrative of an event, which exists in two contested forms (those of the accusers and the accused);
2. Sufficient points of disagreement for us to be able to pose varied, though related questions, in counterbalanced designs over the course of a brain imaging study (thereby reducing the potential for automation of the subject's responses, which might occur if one question were asked continuously);
3. A subject's account that is not weakened by what we might term 'narrative in-homogeneity'. If there are 12 key points to the narrative, the statistical power of our paradigm would be reduced if, for instance, three were falsifications; and
4. The events concerned should be describable in unambiguous terms (requiring the subject to address a single, specific episode).

In a recent study (Spence, Kaylor-Hughes, Brook, Lankappa, & Wilkinson, 2007) we examined a woman who was accused of harming a child. We adapted the technique so that she might be scanned four times, using a bank of 36 questions, administered in counterbalanced designs, while we simultaneously recorded motor responses. Details of the case were obtained on interview and using trial-related transcripts. We invited the subject to respond to questions under two conditions:

1. Giving what she would regard as the 'truth'; and
2. Giving what she would regard as a 'lie' response (endorsing the view of her accusers).

We predicted that if the subject was essentially truthful, then there would be greater activation in VLPFC and longer response times (RTs) when she endorsed the views of her accusers (i.e., when she 'lied'). We posited that the opposite finding (VLPFC activation and longer RTs exhibited when she endorsed her own account) would be inconsistent with her claims of innocence. From a purely statistical perspective the null hypothesis is most likely to be borne out in such experiments, i.e., there is a 95% probability that no discernible difference will be found between brain maps or response times under each condition. We have to state that we cannot prove guilt or innocence. We are merely asking whether brain scans and response times are 'consistent' or 'inconsistent' with the subject's account. In this case, we were able to obtain clear-cut results, which suggest that the protocol has potential. When the subject endorsed the accusations against her (i.e., when she 'admitted guilt') her prefrontal activity and RTs were indeed significantly increased (Figure 3). In this subject, admitting guilt resembled lying. Hence, though we have not proven her innocence, we have demonstrated that her behavioural and functional anatomical parameters behave 'as if' she were wrongly convicted.

In developing this line of work, we have generated future hypotheses on the basis of past experiments where we were able to determine which 'half' of the responses were definitely lies, either through the use of screening interviews prior to the procedure (as in Spence et al., 2001) or concealed testimonies (set aside during scanning), later decoded (post-scanning; as in Spence et al., 2004). So long as there is a way of establishing 'ground

Areas where 'Lies' activate more than 'Truths'

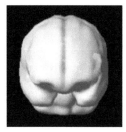

Figure 3. Images obtained from a subject who professes her innocence, the result of the 'lie-truth' cognitive subtraction when she answered questions regarding an alleged offence (Spence et al., 2007). The subject exhibits greater activity in a number of prefrontal regions when endorsing the account of her accusers.

truth', then there is a means of discovering the neural correlates of lying. However, once we enter the forensic arena we are unlikely to know the ground truth and so the logic of the experiment is reversed:

> We go from asking 'what are the correlates of (known) lies?' to asking 'do these (predicted) neural patterns denote (unknown) lies'? The latter question is far more difficult to answer.

Searching for replications

However, before we go too far along the road to forensic fMRI we must step back, to consider another serious question: can fMRI yield reliable data in this context; in other words, can investigators replicate their own and others' findings? This is a difficult subject to address definitively, given that the first such study was only published in 2001 (Spence et al., 2001, above), and that there are few strictly comparable papers to date. Indeed, it is possible, if playing 'Devil's Advocate' (Spence, 2008), to paint a very negative view of this enterprise. We think the field is mixed at the moment: there are some similarities across studies; however, there are also some surprising inconsistencies (below).

While some authors have been particularly interested in studying subjects while they tell lies about their *pasts* (Abe et al., 2005; Abe, Suzuki, Mori, Itoh, & Fujii, 2007; Ganis, 2003; Lee et al., 2002, 2005; Nunez, Casey, Egner, Hare, & Hirsch, 2005; Spence et al., 2001, 2004), others have been more concerned with lies about the *present* e.g., through scenarios played out during or immediately prior to scanning (Davatzikos et al., 2005; Kozel et al., 2004a, 2005; Kozel, Kozel, Padgett, & George, 2004b; Langleben et al. 2002, 2005; Mohammed, 2006; Phan et al., 2005). Consistent with our own model (Figure 1) most authors have demonstrated increased activation of prefrontal regions during lying, although the foci implicated have varied between ventrolateral prefrontal (Abe et al., 2005; Kozel et al., 2004a, 2004b, 2005; Lee et al., 2005; Nunez et al., 2005; Phan et al., 2005; Spence et al., 2001, 2004), anterior cingulate (Abe et al., 2005; Ganis, Kosslyn, Stose, Thompson, & Yurgelun-Todd, 2003; Kozel et al., 2004a, 2004b, 2005; Langleben et al., 2002; Lee et al., 2005; Nunez et al., 2005; Spence et al., 2001) and dorsolateral prefrontal cortices (Abe et al., 2005, 2007; Ganis et al., 2003; Lee et al., 2002, 2005; Nunez et al., 2005; Phan

et al., 2005). Also consistent with the model is the finding that all except one study to date (Langleben et al., 2005) have failed to find areas in the brain that exhibit greater activation during truth-telling (compared with lying); suggesting that 'truthfulness' comprises a relative baseline in human communication; a response tendency that must be suppressed if the liar is to 'succeed' in their deceit (Spence, 2004).

Nevertheless, when looked at another way, the data are not so compelling. The current fMRI literature comprises 16 peer-reviewed data papers relating to deception, emerging from 10 groups of authors, with multiple publications arising from four: Langleben and colleagues (three papers), Kozel and colleagues (also three), Lee and colleagues (two), and our group (two). As the groups have used different paradigms (above) it would be surprising if all studies revealed the same findings yet we should expect to see internal replications (within groups).

Are the findings consistent when individual groups have used the same method twice? Has anyone managed to replicate their own study's findings? This has been something of a problem.

(a) The Guilty Knowledge Test (GKT) studies. If we begin with the study reported by Langleben's group in 2002, using the GKT, adapted from the polygraph literature, then we can see that this early study discovered discrete neural correlates of deceptive responding. The GKT requires the subject to withhold knowledge of a datum that is only known to them (apparently); in this case, the identity of a playing card. When subjects withheld such knowledge in this study they exhibited activation of anterior cingulate cortex (ACC) and a swathe of left sensorimotor cortices (including the left inferior parietal lobule, Brodmann Area 40; Langleben et al., 2002). These findings are consistent with the cognitive conflict involved in withholding such a pre-potent response (the card's identity), and we might infer that predominantly motor mechanisms are involved in deception in this context (where a manual response is required, through the pressing of a button; e.g., responding 'no' instead of 'yes').

However, complications arise when we search for replications. Phan and colleagues (2005) used the GKT paradigm and found different frontal regions to be activated during lies, though they replicated activation of left parietal cortices (BA 40). In a recent study, Gamer, Bauermann, Stoeter, and Vossel (2007) applied the GKT and found further frontal foci to be activated during deception, though they did replicate the left parietal finding (in BA 40). So, in contrast to most authors' *a priori* hypotheses (and our model) it is left parietal cortex that appears to provide the most replicable feature of deception practiced during the GKT (Gamer et al., 2007; Langleben et al., 2002; Phan et al., 2005).

However, further complications arise when we examine the Langleben group's attempts to replicate their own findings. In their 2005 papers (Davatzikos et al., 2005; Langleben et al., 2005), where the same data are analysed in two different ways, we find that:

1. The initial ACC and inferior parietal lobe findings are not replicated at sufficient statistical significance to be reported by most neuroimaging groups;
2. The parietal lobes (bilaterally) now exhibit greater activity during truthful responding than withholding the truth; and
3. The behavioural and functional anatomical findings are reported differently across papers: while truthful RTs appear only qualitatively longer than lie responses in Langleben et al. (2005), they are reported as significantly longer in Davazitkos et al. (2005); while very many regions are more active during truthful responding in Langleben et al. (2005), a different subset of foci are maximally informative in Davazitkos et al. (2005; admittedly, following the application of a different, novel, analytic method).

So, if we were to attempt to diagnose deception (at this moment) on the basis of the GKT paradigm, as applied in fMRI, we would not be able to rely upon the frontal signals reported by different authors (replications are lacking) and the consistency evinced with respect to left parietal cortex appears perplexing: it is more active during deception in some studies (Gamer et al., 2007; Langleben et al., 2002; Phan et al., 2005); more active during truth-telling in others (Langleben et al., 2005).

(b) Mock Crime Scenario Studies. Are things any better among the mock-crime studies? Kozel's group has contributed most here: they have reported similar studies on three occasions (Kozel et al., 2004a, 2004b, 2005). In each paper they report predominantly right-sided frontal activation during deception compared with truthful responding (as indicated through manual movements). They go on to identify certain foci as being of particular importance and then count the voxels activated in these areas among subsequent subjects, in an attempt to diagnose deception. Like the Langleben group, Kozel's group has tested this approach rigorously but, again, there are inconsistencies. If we examine the foci activated across the three studies it is clear that though there may be a relative preponderance of right-sided frontal lobe activations during deceptive responding the precise foci implicated vary across reports: from orbitofrontal (Kozel et al., 2004a) to ACC and more posterior frontal regions (Kozel et al., 2004b) to ACC and dorsal (superior) prefrontal regions (Kozel et al., 2005). These areas of the frontal lobe are quite distinct: they have different cytoarchitectures, are implicated in different cognitive functions and, though they may co-activate during a variety of cognitive tasks (Duncan & Owen, 2000), we must acknowledge that Kozel and colleagues' informative regions, the areas they implicate in deception, have moved across the right frontal lobe.

Unfortunately, the other mock crime-fMRI study is uninformative: Mohamed et al. (2006) do not report a formal contrast (statistical comparison) between deceptive and truthful responding, so we cannot tell which, if any, of Kozel's findings they might have replicated.

(c) Feigning Memory Impairment. Lee et al. (2002, 2005) have twice reported studies of subjects engaged in malingered memory impairment, during which they were required to 'pretend' to be impaired, but not *too* impaired; because if they performed worse than chance then they might be 'found out'. In each of their studies they found widespread activation among frontal and posterior cortical regions, greater during malingered inaccuracy than truthful, accurate, responding. Does their second set of findings constitute a replication of their first?

To begin with, it is important to note that the second experiment is not identical to the first: the memory domains combined to produce the first set of results related to (long term) autobiographical and (short term) digit memory (the findings are similar, though not identical under both of these conditions; Lee et al., 2002). In the second study the authors probed (short term) verbal and digit memory and the data for these two forms of memory were not presented separately but as a 'conjunction', i.e., a combined analysis (Lee et al., 2005). The areas of activation common to both papers (during deceptive responding) are bilateral dorsolateral prefrontal (BA 9), left anterior prefrontal (BA 10), right dorsolateral prefrontal (BA 46) and bilateral premotor (BA 6) cortices. However, to identify such apparent congruence one must combine data from different experiments, probing different domains of memory, incorporating findings that are analysed using different statistical approaches.

Hence, we should reiterate that, while there are commonalities across studies and centres, so far, most authors have either not attempted or not succeeded in precisely replicating their own findings. Our model is found to be supported at a conceptual level (in that greater prefrontal activity is associated with deceit in nearly all studies reported and truth associated with greater prefrontal activation in only one). Nevertheless, at a detailed, anatomical level of analysis the model is found wanting (investigators often fail to replicate their own findings). In our defence, we have replicated the finding of VLPFC activation associated with deception (though our method has not received attempted replication at other laboratories, to our knowledge).

CAVEATS AND CONCERNS: IS THERE AN ETHICAL WAY FORWARD?

If we set aside our concerns about reliability, and assume that such technical issues can be 'ironed out', there are still 2 further questions that we must ask concerning the application of fMRI to lie detection:

1. *Can* we do it?
2. *Should* we do it?

The first of these is a technical question, the second ethical.

We have shown that the technical question awaits formal resolution, though it is possible to see how this might be achieved. We require more prospective studies, where those 'reading' the scans do not already know the 'ground truth' of their subjects (although it might be known to adjudicators). In other words, the first question seems tractable. Yet, it is also more complex than at first meets the eye: for the question of detecting deception through scanning is not merely a question of neuroimaging methodology; it implicates an entire process, stretching from the elicitation of information regarding the 'crime',

through the construction and delivery of unambiguous questions concerning the 'truth', and the correct attribution of meaning to the images seen (i.e., prefrontal activation indicates deception and not some other, confounding cognitive process). Much more work is required to dissect the stages of this process and to compare different laboratories' success when faced with the same forensic question. The scanning protocol only arises close to the end of the process.

The second question ('should we do it?') is one for society at large and it would be presumptuous of us to proffer conclusions. However, we might offer a preliminary thought: if one is concerned with medical ethics and the proper treatment of human subjects, then it is difficult to foresee a situation in which it would be justifiable to scan people against their will. Hence, the special case of an alleged miscarriage of justice may provide the appropriate medium for ongoing work, in the near-term future, so long as subjects are requesting investigation; though we stress subjects should be 'truly' free to volunteer (and not exposed to subtle forms of coercion).

DOES THIS WORK TELL US ANYTHING ABOUT OURSELVES?

We wish to close by asking a broader question, provoked by this line of scientific work. Does the scanning of lying tell us anything 'extra'?

Does biology 'represent' the inauthentic?

One dataset that we have obtained has suggested a new line of thought to us, and though it is a preliminary finding we think it of sufficient interest to consider here, as it exemplifies how the study of deception can throw up insights into ourselves that we may not have been expecting.

In a study of vocally expressed lies (Spence et al., 2004), we asked subjects to provide confidential details of two true scenarios from their lives that most people would regard as embarrassing. Details were kept in sealed envelopes throughout the study and subjects were presented with aural questions when scanned using fMRI. However, as well as wishing to contrast lies and truths we were interested in comparing the same subjects under two further cognitive conditions: when they obeyed the request of their interlocutor to repeat a word, and when they declined to do so (saying the word 'no'). Stimulus and scanning parameters were matched across four conditions: telling the truth, lying, complying with their interlocutor and defying him (saying no).

While we anticipated the relative activation of bilateral VLPFC in the lying condition (cf. 'truth'), we did not expect this profile to be shared by the compliance condition (Figure 4, right). What we found was that VLPFC activation increased with deceptive and compliant responding and decreased during truthfulness and defiance.

This is only one experiment and we do not wish to over-interpret our data but we might ask: what is shared by lies and compliance and what is common to truth and defiance? It occurs to us that there is at least one, rather intriguing, hypothesis: that VLPFC is, in some way, involved in the enactment of inauthentic responses; whereas it is not activated when subjects tell the truth or defy their interlocutor. In this way, our data seem to shed light upon the possible pathophysiology of the 'pseudopsychopathic syndrome' described by Blumer and Benson (1975), where those suffering OFC lesions become 'brash', 'outspoken', lacking in 'tact'. Such pathological findings, and our more localized findings in healthy subjects argue for a confluence of deception with obedience, and honesty with defiance. This exemplifies how work in this area may provoke us towards possibly unexpected hypotheses regarding the human brain (and moral action).

CONCLUSION

We began by stating that the subjective state is pivotal to apportioning responsibility for antisocial acts. We rehearsed the evidence that knowing deception is a task likely to be reliant upon prefrontal executive systems and we went on to show, that among the majority of functional neuroimaging studies investigating deception, prefrontal activity has been detectable (although somewhat unreliable). Nevertheless, while such studies have mostly detected the correlates of deception, using protocols where 'ground truth' has been known to investigators, the central problem posed by forensic application is a reversal of this logic: it is to discover deception on the basis of brain activity (without recourse to ground truth). Obviously, this is a much more difficult proposition. We have commented upon one method by which this might be approached and we have indicated that orbitofrontal cortex in general, and VLPFC in particular, continues to be a candidate region of especial interest in the

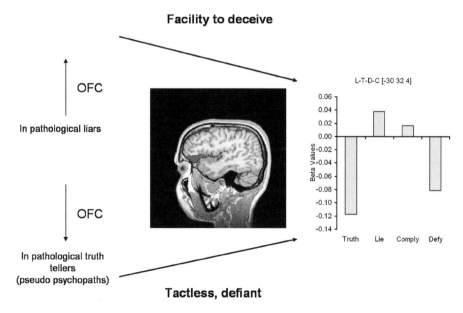

Figure 4. A cartoon demonstrating the convergence of deception-related data upon orbitofrontal cortices in humans. Left: increased OFC white matter is associated with pathological lying in antisocial people (Yang et al., 2007); lesions of OFC are associated with outspokenness, tactlessness and inappropriate sharing of intimacies (Blumer & Benson, 1975; Beer et al., 2003). In our data from a vocal lying study (Spence et al., 2008), the response of ventrolateral prefrontal cortex is modulated according to what seems to be the authenticity or otherwise of the subjects' responses: increased activity during lies and compliance, reduced activity during truths and defiance (right).

physiology of deception and the pathophysiology of pathological lying and truthfulness.

REFERENCES

Abe, N., Suzuki, M., Tsukiura, T., Mori, E., Yamaguchi, K., Itoh, M., & Fujii, T. (2006). Dissociable roles of prefrontal and anterior cingulate cortices in deception. *Cerebral Cortex, 16*, 192–199.

Abe, N., Suzuki, M., Mori, E., Itoh, M., & Fujii, T. (2007). Deceiving others: Distinct neural responses of the prefrontal cortex and amygdala in simple fabrication and deception with social interactions. *Journal of Cognitive Neuroscience, 19*, 287–295.

Bass, C. (2001). Factitious disorders and malingering. In P. W. Halligan, C. Bass, & J. C. Marshall (Eds), *Contemporary approaches to the study of hysteria: Clinical and theoretical perspectives* (pp. 126–142). Oxford: Oxford University Press.

Beer, J. S., Heerey, E. A., Keltner, D., Scabini, D., & Knight, R. T. (2003). The regulatory function of self-conscious emotion: Insights from patients with orbitofrontal damage. *Journal of Personality and Social Psychology, 85*, 594–604.

Blumer, D., & Benson, D. F. (1975). Personality changes with frontal and temporal lobe lesions. In D. F. Benson & D. Blumer (Eds), *Psychiatric aspects of neurologic disease* (pp. 151–170). New York: Grune & Stratton.

Butters, N., Butter, C., Rosen, J., & Stein, D. (1973). Behavioural effects of sequential and one-stage ablations of orbitofrontal prefrontal cortex in the monkey. *Experimental Neurology, 39*, 204–214.

Byrne, R. W. (2003). Tracing the evolutionary path of cognition. In M. Brune, H. Ribbert, & W. Schiefenhovel (Eds), *The social brain: Evolution and pathology* (pp. 43–60). Chichester, UK: Wiley.

Byrne, R. W., & Corp, N. (2004). Neocortex size predicts deception rate in primates. *Proceedings of the Royal Society of London, series B, 271*, 1693–1699.

Davatzikos, C., Ruparel, K., Fan, Y., Shen, D. G., Acharyya, M., Loughead, J. W., Gur, R. C., & Langleben, D. D. (2005). Classifying spatial patterns of brain activity with machine learning methods: Application to lie detection. *NeuroImage, 28*, 663–668.

Duncan, J., & Owen, A. M. (2000). Common regions of the human frontal lobe recruited by diverse cognitive demands. *Trends in Neurosciences, 23*, 475–483.

Ekman, P., & O'Sullivan, M. (1991). Who can catch a liar? *American Psychologist, 46*, 913–920.

Fletcher, P. C., Happe, F., Frith, U., Baker, S. C., Dolan, R. J., Frackowiak, R. S. J., & Frith, C. D. (1995). Other minds in the brain: A functional imaging study of 'theory of mind' in story comprehension. *Cognition, 57*, 109–128.

Ford, C. V., King, B. H., & Hollender, M. H. (1988). Lies and liars: Psychiatric aspects of prevarication. *American Journal of Psychiatry, 145*, 554–562.

Galbraith, J. K. (2004). *The economics of innocent fraud.* London: Penguin.

Gamer, M., Bauermann, T., Stoeter, P., & Vossel, G. (2007). Covariations among fMRI, skin conductance and behavioural data during processing of concealed information. *Human Brain Mapping, 28*, 1287–1301.

Ganis, G., Kosslyn, S. M., Stose, S., Thompson, W. L., & Yurgelun-Todd, D. A. (2003). Neural correlates of different types of deception: An fMRI investigation. *Cerebral Cortex, 13*, 830–836.

Giannetti, E. (2000). *Lies we live by: The art of self-deception* (translated by J. Gledson). London: Bloomsbury.

Granhag, P. A., & Stromwall, L. A. (2004). *The detection of deception in forensic contexts.* Cambridge: Cambridge University Press.

Happe, F. (1994). *Autism: An introduction to psychological theory.* Hove, UK: Psychology Press.

Hughes, C. J., Farrow, T. F. D., Hopwood, A.-C., Pratt, A., Hunter, M. D., & Spence, S. A. (2005). Recent developments in deception research. *Current Psychiatry Reviews, 1*, 271–279.

Iversen, S. D., & Mishkin, M. (1970). Perseverative interference in monkeys following selective lesions of the inferior prefrontal convexity. *Experimental Brain Research, 11*, 376–386.

Kozel, F. A., Revell, L. J., Lorberbaum, J. P., Shastri, A., Elhai, J. D., Horner, M. D., Smith, A., Nahas, Z., Bohning, D. E., & George, M. S. (2004a). A pilot study of functional magnetic resonance imaging brain correlates of deception in healthy young men. *Journal of Neuropsychiatry and Clinical Neurosciences, 16*, 295–305.

Kozel, F. A., Padgett, T. M., & George, M. S. (2004b). A replication study of the neural correlates of deception. *Behavioural Neuroscience, 118*, 852–856.

Kozel, F. A., Johnson, K. A., Mu, Q., Grenesko, E. L., Laken, S. J., & George, M. S. (2005). Detecting deception using functional magnetic resonance imaging. *Biological Psychiatry, 58*, 605–613.

Langleben, D. D., Schroeder, L., Maldjian, J. A., Gur, R. C., McDonald, S., Ragland, J. D., O'Brien, C. P., & Childress, A. R. (2002). Brain activity during simulated deception: An event-related functional magnetic resonance study. *NeuroImage, 15*, 727–732.

Langleben, D. D., Loughead, J. W., Bilker, W. B., Ruparel, K., Childress, A. R., Busch, S. I., & Gur, R. C. (2005). Telling truth from lie in individual subjects with fast event-related fMRI. *Human Brain Mapping, 26*, 262–272.

Layard, R. (2005). *Happiness: Lessons from a new science.* London: Allen Lane.

Lee, T. M. C., Liu, H.-L., Tan, L.-H., Chan, C. C. H., Mahankali, S., Feng, C.-M., Hou, J., Fox, P. T., & Gao, J.-H. (2002). Lie detection by functional magnetic resonance imaging. *Human Brain Mapping, 15*, 157–164.

Lee, T. M. C., Liu, H.-L., Chan, C. C. H., Ng, Y.-B., Fox, P. T., Gao, J.-H. (2005). Neural correlates of feigned memory impairment. *NeuroImage, 28*, 305–313.

Machiavelli, N. (1999). *The Prince* (translated by G. Bull). London: Penguin (first published 1961).

Mealey, L. (1995). The socio-biology of sociopathy: An integrated evolutionary model. *Behavioural and Brain Sciences, 18*, 523–599.

Mohamed, F. B., Faro, S. H., Gordon, N. J., Platek, S. M., Ahmad, H., & Williams, J. M. (2006). Brain mapping of deception and truth telling about an ecologically valid situation: Functional MR imaging and polygraph investigation – initial experience. *Radiology*, *238*, 679–688.

Nunez, J. M., Casey, B. J., Egner, T., Hare, T., & Hirsch, J. (2005). Intentional false responding shares neural substrates with response conflict and cognitive control. *NeuroImage*, *25*, 267–277.

Oborne, P. (2006). *The use and abuse of terror: The construction of a false narrative on the domestic terror threat*. London: Centre for Policy Studies.

O'Connell, S. (1998). *Mindreading: An investigation into how we learn to love and lie.* London: Arrow Books.

Phan, K. L., Magalhaes, A., Ziemlewicz, T. J., Fitzgerald, D. A., Green, C., & Smith, W. (2005). Neural correlates of telling lies: A functional magnetic resonance imaging study at 4 Tesla. *Academic Radiology*, *12*, 164–172.

Roget's Thesaurus (2004 edition). London: Penguin.

Shallice, T. (1988). *From neuropsychology to mental structure*. Cambridge: Cambridge University Press.

Sokol, D. K. (2006). How the doctor's nose has shortened over time: A historical overview of the truth-telling debate in the doctor-patient relationship. *Journal of the Royal Society of Medicine*, *99*, 632–636.

Spence, S. A. (2004). The deceptive brain. *Journal of the Royal Society of Medicine*, *97*, 6–9.

Spence, S. A. (2005). Prefrontal white matter – the tissue of lies? Invited commentary on . . . Prefrontal white matter in pathological liars. *British Journal of Psychiatry*, *187*, 326–327.

Spence, S. A. (2008). Playing Devil's Advocate: The case against fMRI lie detection. *Legal and Criminological Psychology*, *13*, 11–25.

Spence, S. A., Farrow, T. F. D., Herford, A. E., Wilkinson, I. D., Zheng, Y., & Woodruff, P. W. R. (2001). Behavioural and functional anatomical correlates of deception in humans. *NeuroReport*, *12*, 2849–2853.

Spence, S. A., Hunter, M. D., & Harper, G. (2002). Neuroscience and the will. *Current Opinion in Psychiatry*, *15*, 519–526.

Spence, S. A., Hunter, M. D., Farrow, T. F. D., Green, R. D., Leung, D. H., Hughes, C. J., & Ganesan, V. (2004). A cognitive neurobiological account of deception: Evidence from functional neuroimaging. *Philosophical Transactions of the Royal Society of London series B*, *359*, 1755–1762.

Spence, S. A., Kaylor-Hughes, C. J., Brook, M. L., Lankappa, S. T., & Wilkinson, I. D. (2007). 'Munchausen's syndrome by proxy' or a 'miscarriage of justice'? An initial application of functional neuroimaging to the question of guilt versus innocence. *European Psychiatry,* published online 25th October, 2007.

Spence, S. A., Kaylor-Hughes, C. J., Farrow, T. F., & Wilkinson, I. D. (2008). Speaking of secrets and lies: The contribution of ventrolateral prefrontal cortex to vocal deception. *NeuroImage*, *40*, 1411–1418.

Starkstein, S. E., & Robinson, R. G. (1997). Mechanism of disinhibition after brain lesions. *Journal of Nervous and Mental Diseases*, *185*, 108–114.

Vrij, A. (2000). *Detecting lies and deceit: The psychology of lying and the implications for professional practice*. Chichester, UK: Wiley.

Wilkinson, R. G. (2005). *The impact of inequality: How to make sick societies healthier*. London: Routledge.

Yang, Y., Raine, A., Lencz, T., Bihrle, S., LaCasse, L., & Colletti, P. (2005). Prefrontal white matter in pathological liars. *British Journal of Psychiatry*, *187*, 320–325.

Yang, Y., Raine, A., Narr, K. L., Lencz, T., LaCasse, L., Colletti, P., & Toga, A. W. (2007). Localisation of increased prefrontal white matter in pathological liars. *British Journal of Psychiatry*, *190*, 174–175.

NEUROCASE
2008, 14 (1), 82–92

Detecting concealed information using brain-imaging technology

Mart Bles[1] and John-Dylan Haynes[1,2]

[1]Bernstein Center for Computational Neuroscience Berlin, Charité – Universitätsmedizin Berlin, Germany
[2]Max Planck Institute for Cognitive and Brain Sciences, Leipzig, Germany

Many conventional techniques for revealing concealed information have focused on detecting whether a person is responding truthfully to specific questions, typically using some form of lie detector. However, lie detection has faced a number of criticisms and it is still unclear to what degree conventional lie detectors can be used to reveal concealed knowledge in applied real-world settings. Here, we review the key problems with conventional lie-detection technology and critically discuss the potential of novel techniques that aim to directly read concealed mental states out of patterns of brain activity.

Keywords: Lie detection; Deception; Concealed information; Brain imaging; fMRI; Decoding; Multivariate analysis.

INTRODUCTION

In many everyday social interactions, humans have to judge what is on another person's mind. This ability is highly limited, however, which becomes obvious when persons deliberately conceal their thoughts. Until recently, forensic technologies for detecting concealed mental states and knowledge have focused on measuring changes in overall arousal or overall brain activity to determine whether people are telling the truth in response to specific questions. It has become clear though, that the use of lie detection to uncover concealed information is very difficult. Recently, a more promising technique has emerged that aims to directly reveal if a person exhibits particular brain responses when confronted with crime-related *information* that could only be known to the perpetrator. Such developments have led to speculations about the development of a technology that could directly read out the contents of another person's mind, including their memories and intentions. Here, we will review the difficulties encountered in classical approaches for detecting concealed knowledge using lie-detection and will discuss the perspectives and key challenges for methods aimed at directly detecting concealed mental states from a person's brain activity.

CONVENTIONAL LIE-DETECTION

The conventional way to reveal whether an interviewee is concealing information is by assessing whether the answers they give to specific questions are truthful or deceptive. The most obvious and historically earliest way to detect if someone is lying is to make use of behavioural cues. Many people hold the belief that liars can be unmasked by 'nervous behaviour' and altered eye contact (DePaulo, 1992) and researchers have long debated

This work was funded by the Dutch Science Foundation (NWO), the German Federal Ministry of Education and Research (BMBF) and by the Max Planck Society.

Address correspondence to Mart Bles or John-Dylan Haynes, Bernstein Center for Computational Neuroscience, Charité – Universitätsmedizin Berlin, Haus 6, Philippstrasse 13, 10115 Berlin, Germany (E-mail: mart.bles@bccn-berlin.de; haynes@bccn-berlin.de).

DOI: 10.1080/13554790801992784

whether attempts to deceive could be given away by changes in body posture, eye contact, language fluency or brief changes in facial expression called microexpressions (see Vrij & Mann, 2001, for a review). Changes in these measures are thought to be signs of physiological arousal caused by feelings of guilt or anxiety during deception, increased cognitive demands associated with concealing information, or increased efforts of behavioural control in order to prevent being caught. Contrary to popular belief, however, only well-trained experts can interpret these behavioural cues above chance, while laymen are surprisingly poor at detecting lies (Ekman & O'Sullivan, 1991).

A more direct approach to assess if someone is telling the truth is to measure physiological signs of peripheral arousal. One of the oldest methods originates from China, in around 1000 BC, where suspects were asked to fill their mouths with dry rice and then spit it out again. If the rice was still dry, the lack of salivation could indicate fear and anxiety, and was interpreted as a sign of lying (Ford, 2006). More recently, methods such as thermal body imaging (Pavlidis, Eberhardt, & Levine, 2002) or voice-stress analysis (Horvath, 1982) have been developed, although the validity of the latter has been questioned (see Gamer, Rill, Vossel, & Godert, 2006). For decades, the standard way to detect deception based on physiological measures has been the so-called 'polygraph test'. The polygraph simultaneously measures multiple physiological parameters that indicate arousal such as heart rate, respiration rate and galvanic skin response. In most cases, the polygraph is used in combination with a specific questioning strategy known as the Control Question Test (CQT). This test compares changes in peripheral arousal to selected categories of questions. *Irrelevant questions* concern information that should not trigger a strong emotional response (e.g., 'Were you born in 1973?'). *Control questions* probe minor misconducts committed by almost everyone and are designed to trigger an emotional response (e.g., 'Did you ever steal anything?'). *Relevant questions* concern the matter under investigation (e.g., 'Did you steal that car?'). If the physiological response to relevant questions is larger than to control questions, the suspect is assumed to be lying.

Polygraphy based on the CQT, however, has a number of serious shortcomings. For example, the threshold for classifying a response as lie or truth crucially depends on the choice of control questions. In addition, the physiological measures on which polygraph tests are based are also sensitive to other factors such as excitement, fear and anger (Steinbrook, 1992). The questioned person, for example, may know which question is relevant (e.g., 'Did you kill your wife?') and might respond to it with increased arousal, which could lead to a high percentage of false positives. On the other hand, in the population for which these tests are intended (i.e., criminals), psychopathy and antisociality are overrepresented, and these traits may lead to autonomic underarousal (for a review, see Lorber, 2004). This could reduce the sensitivity of polygraphy-based questioning techniques to uncover liars in this population (but see Verschuere, Crombez, Koster, & De Clercq, 2007). To make things worse, it is possible to avoid detection by learning to control the physiological reactions in response to the questions (Honts, Raskin, & Kircher, 1994). Detailed descriptions of countermeasures can easily be obtained from web pages that train people to trick polygraph tests. Finally, scientific validation of the CQT is poor, and the questioning method is not standardised (Ben-Shakhar, 2002; Furedy & Heslegrave, 1991). For all these reasons, polygraphy has been widely criticised in the scientific community. A recent review by the Board on Behavioral, Cognitive, and Sensory Sciences and Education (BCCSE, 2003) estimated the median accuracy index for 52 scientifically sound studies to be around .86, and concluded that polygraph testing is too inaccurate to be used in practical settings and that alternative methods ought to be investigated.

BRAIN-BASED LIE DETECTION

One potential alternative is to detect deception by directly observing the neural correlates of lying in the brain. This is possible using neuroimaging techniques such as PET and fMRI. These methods measure local increases in oxygen consumption or blood flow and thus reflect brain activity at a high spatial precision. They have been shown to be highly sensitive to a large number of cognitive processes, including those relevant for deception such as memory recall (Burgess, Maguire, & O'Keefe, 2002; Milner, Squire, & Kandel, 1998), cognitive control (Botvinick, Nystrom, Fissell, Carter, & Cohen, 1999) and motivation (Elliott, Friston, & Dolan, 2000). The first neuroimaging studies of deception were not aimed at developing a lie-detection method, but at revealing the neural processes involved in lying. Thus, no trial-by-trial assessments of deception were made. Instead they investigated the neural mechanisms of the cognitive operations involved in deception during blocks (or sequences) of a few 'deception' trials versus blocks of a few 'truth' trials. In a typical paradigm, participants were asked simple questions about episodic (e.g., 'Who was your best friend at primary school?') (Abe et al., 2006; Abe, Suzuki, Mori, Itoh, & Fujii, 2007; Ganis, Kosslyn, Stose, Thompson, & Yurgelun-Todd, 2003; Nunez, Casey, Egner, Hare, & Hirsch, 2005; Spence et al., 2001) or declarative memory (e.g., 'Does a bicycle have six wheels?') (Nunez et al., 2005). They were asked to respond truthfully in some blocks and deceptively in others, regardless of the content of the question. Brain activations in response to truthful answers were compared with those to deceptive ones to expose areas involved in generating deceptive responses.

Although the paradigms employed vary considerably, one general impression is confirmed by these studies: lying involves more effort than telling the truth (see Spence et al., 2004). Several brain areas respond more strongly when people tell lies. Most notably, increased activity in ventrolateral (VLPFC) and dorsolateral prefrontal cortex (DLPFC) and anterior cingulate cortex (ACC) has been reported during lying. Interestingly, VLPFC has previously been implied to be involved in response inhibition (Starkstein & Robinson, 1997), which is in accordance with the idea that lying involves inhibiting pre-potent responses (Spence et al., 2001, 2004). Response inhibition may also activate ACC (Botvinick et al., 1999; Botvinick, Cohen, & Carter, 2004), as does signalling the presence of response or processing conflicts (Badre & Wagner, 2004; MacDonald, Cohen, Stenger, & Carter, 2000; Ruff, Woodward, Laurens, & Liddle, 2001). Finally, DLPFC has been associated with working memory load (Courtney, Petit, Haxby, & Ungerleider, 1998; Smith & Jonides, 1999), goal maintenance (MacDonald et al., 2000), response selection (Rowe & Passingham, 2001), and response generation (Frith, Friston, Liddle, & Frackowiak, 1991). These findings suggest a strong role for prefrontal cortex in the inhibition of truthful responses and the generation of false responses that constitute the lie (Spence et al., 2004). No areas have been reported that are reliably more active during truth-telling than during lying.

Due to the blocked presentation, these studies were not able to assess whether a suspect was lying about a specific single item of information. More recent research paradigms have been developed in which truthful and deceptive responses are not distributed in blocks, but vary on a trial-by-trial basis and thus can be used to test for the truth of answers to selected questions. These paradigms employed a variety of tasks in which subjects lied about the identity of objects in a mock crime (Kozel et al., 2005) or about the identity of playing cards they had been shown (Gamer, Bauermann, Stoeter, & Vossel, 2007; Langleben et al., 2002, 2005; Phan et al., 2005). During scanning, participants were asked to press buttons to answer 'yes' or 'no' and to conceal their information by lying in response to specific questions. Since the truthfulness of responses varied from trial to trial, it was possible to measure brain activity in response to specific pieces of information. Importantly, the areas that were activated by these paradigms correspond to the ones activated in the block-based paradigms. Phan et al. (2005), for example, observed increased activations in VLPFC and DLPFC, amongst other areas, when participants lied about the identity of a playing card.

Classical neuroimaging studies use a massively univariate approach that involves independently performing the same statistical test at thousands of voxels simultaneously (see Figure 1a). Such an approach has become the standard to investigate brain areas responding differentially to certain mental states, and it has proven useful in studying the neural correlates of deception. However, it does not provide a suitable framework for detecting lies because it does not allow for the assessment of the full amount of information that can be extracted from the entire brain and of the reliability with which a deceptive response can be identified. A more powerful approach is to accumulate information across multiple brain areas, which increases the amount of information that can be extracted from brain activity (Kozel et al., 2005; Langleben et al., 2005). Kozel et al. (2005), for example, had people steal one of two objects in a mock crime. In the subsequent questioning session, subjects denied taking either object while their brain activity was being monitored. A statistical model that combined the activation of several cortical regions was able to predict whether a subject's response was truthful or deceptive. The model could differentiate between 'truth' and 'lie' for more than 90% of the subjects, even when tested on an independent group of subjects.

Another way of accumulating information across brain regions is the use of multivariate

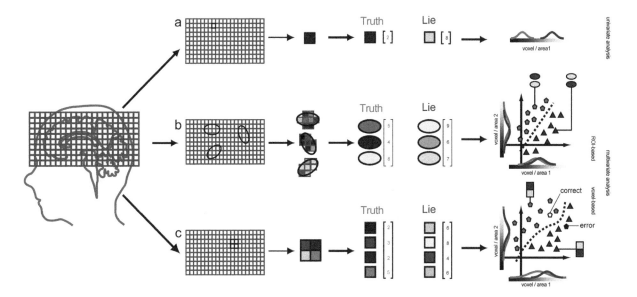

Figure 1. Pattern classification for detection of deception and concealed knowledge. (a) This shows a hypothetical voxel that responds strongly when a person is lying (blue) and weakly when they are telling the truth (red). In this case, the decision whether the person is lying or not could be made based on the activity in this single voxel because the distributions are widely separated (far right). However, in real neuroimaging data, the responses in individual voxels are only slightly separated and thus individual voxels are not sufficient to tell truth from lie. (b) If the neurocognitive correlates of deception are specific, global patterns of brain activity, then pattern recognition can be used to recognise deception by training a classifier to identify the occurrence of this pattern of brain activity. The right-hand figure shows the principle of pattern recognition using the average activity in two different brain regions as an example. The average activation values of the two individual regions can be plotted on the x- and y-axis yielding a number of measurement points, one for each brain pattern. Red points correspond to global patterns acquired during truthful responding and blue points correspond to global patterns acquired during deception. The decision cannot be made based on individual regions but can easily be made when taking into account the combined activation in both regions. Here the decision is made using a linear decision boundary (dashed line). A key question is, however, whether such unique signature patterns exist that are valid for different people, different situations and different types of lies. (c) This shows a local decoding approach that can be used to reveal information stored in micro-patterns of local cortical maps (Haynes & Rees, 2006). The right hand figure shows a case where the decision boundary is non-linear. New measurements of brain activity are classified as lie or truth depending on which side of the boundary they fall (black and white symbols). This figure can be viewed in color on the Journal's website.

pattern recognition (Cox & Savoy, 2003; Haxby et al., 2001). Pattern classification algorithms can learn to distinguish mental states based on the distributed patterns of brain activity associated with those states (see Figure 1b). After the training phase, new brain-activity patterns can be categorised based on their similarity to patterns measured during training (see also Haynes & Rees, 2006). By combining information from activity patterns spanning many regions or voxels, the sensitivity of functional neuroimaging is highly increased. Multivariate techniques may even allow the classification of single fMRI volumes (Haynes & Rees, 2005b). Using such an approach, Davatzikos et al. (2005) trained a classification algorithm to distinguish brain-activity patterns associated with truthful and deceptive responses. Importantly, they were able to classify activity patterns of new participants that had not been used for training the algorithm with about 90% accuracy.

FUNDAMENTAL PROBLEMS WITH LIE DETECTION

Although brain-based detection of deception has initially shown promising results, there are still a number of key challenges. One of the major problems that is shared by brain-based and arousal-based studies on lying is that participants need to be studied during deception scenarios that are as 'real' as possible. In order to accomplish this, some studies have employed the tactic of leading participants to believe that their brain activity or responses are monitored by an external observer who will try to identify deceptive responses (Ganis et al., 2003; Phan et al., 2005; Spence et al., 2001), sometimes combined with the promise of a monetary reward for successful concealment (Kozel, Padgett, & George, 2004; Kozel et al., 2005; Langleben et al., 2002; Mohamed et al., 2006). It remains to be seen, however, whether such manipulations are successful in triggering motivational levels comparable to that of real-life situations, in which peoples' freedom, career or credibility might be lost if they do not convince the interrogators of their innocence (Vrij & Mann, 2001).

A further deviation from real-world situations is that lie detection is mostly investigated in simplified experimental tasks that might not capture the complexity of real-world deception. Participants often respond by pressing one of two buttons corresponding to 'yes' and 'no' answers. This non-verbal mode of communication via an arbitrarily assigned response button is quite far from normal verbal communication. Hence, it is not clear whether the emotional responses associated with spoken lies are similarly present when 'button-press lies' are made. Moreover, if the subject is asked to lie with respect to having seen some objects and to respond truthfully in response to other objects, the task can be thought of as a response remapping problem ('left button when you see this picture versus right button when you see the other picture'), thus the task could in principle be solved without making any reference to the intended meaning of response buttons (Kozel et al., 2004; Langleben et al., 2005; Phan et al., 2005). The finding of prefrontal cortical activation in such deception paradigms might simply indicate the involvement of cognitive control in task switching or response inhibition (Aron, Robbins, & Poldrack, 2004; MacDonald et al., 2000; Miller & Cohen, 2001; van Veen, Cohen, Botvinick, Stenger, & Carter, 2001), possibly independent of their role in deception (see Figure 2). The activation of dorsolateral and ventrolateral prefrontal regions is not limited to situations of deception but is observed in a wide variety of tasks that have been associated with different aspects of 'executive function' (see Duncan & Owen, 2000, and Elliott, 2003, for a review). It remains to be seen which of these neurocognitive processes are intrinsically and necessarily involved in deception

and whether a cortical network can be identified that is specifically involved in the generation of lies.

This obervation directly raises the question of the cognitive operations involved in deception. Acts of deception can be very distinct from one another, so it seems a simplification to even think of deception as a unified class of cognitive mechanisms that can be revealed with a single measurement (Ganis et al., 2003). For example, a deceiver's knowledge of the target's intentions and know-how may enable him to lie more convincingly. So theory of mind (ToM) plays an important role in the interaction between deceiver and target (Spence et al., 2001). More complicated situations may include second-order ToM, like telling the truth when another person thinks it is a lie. Furthermore, motivation plays a key role. People will have a stronger motivation to lie in high-stake situations (e.g., criminal investigations), which has important implications for lie-detection methods (DePaulo, Kashy, Kirkendol, Wyer, & Epstein, 1996; Vrij & Mann, 2001). Thus, a range of cognitive and motivational processes are involved in deception, including memory processes (Ganis et al., 2003; Nunez et al., 2005), strategic decisions (Lee et al., 2002, 2005), theory of mind (Spence et al., 2004), response generation and inhibition (Langleben et al., 2002; Spence et al., 2001), emotional regulation (Abe et al., 2007; Ekman, 1992) and reward expectations (Kozel et al., 2004). Not surprisingly, studies have shown a large variability in the areas reported to be involved in lying. Some studies, for example, do not report activation of DLPFC (e.g., Gamer et al., 2007; Mohamed et al., 2006; Spence et al., 2001), VLPFC (e.g., Abe et al., 2007; Ganis et al., 2003), or ACC (e.g., Gamer et al., 2007; Phan et al., 2005) when participants make a deceptive response. Due to the complexity of different types of deception, it has to be questioned whether a unique set of brain regions can be identified that serve as a reliable indicator of deception per se, rather than the broader class of processes included in 'executive functioning'. This might prove a problem for classification-based techniques of lie detection as well. To make things worse, comprehensive cognitive models of deception have remained illusive up till now.

Finally, fMRI studies on lie detection typically describe young and healthy adults. However, BOLD activity is known to be altered with age (Buckner, Snyder, Sanders, Raichle, & Morris, 2000; D'Esposito, Deouell, & Gazzaley, 2003), in patients with cardiovascular diseases (Pineiro, Pendlebury, Johansen-Berg, & Matthews, 2002;

Figure 2. Similarity between brain patterns characteristic for deception (schematically redrawn from Spence et al., 2001, Figure 1) and brain patterns that occur during response inhibition (schematically redrawn from Blasi et al., 2006, Figure 1b). This figure can be viewed in color on the Journal's website.

Rother et al., 2002), and drug use (Bruhn et al., 1994) or abuse (Levin et al., 1998; Sell et al., 1997). Hence, there is no reason to assume that a method calibrated on young, healthy brains works across age and clinical groups. Similar arguments have been put forward that could limit the generalisation of these techniques to criminal populations (Lykken, 1959).

INDIRECT AND DIRECT READ-OUT OF CONCEALED INFORMATION

Instead of trying to detect lies, an alternative strategy could be to find a way to 'read out' information that is being concealed by the subject. Thus, instead of identifying physiological markers for deception one requires physiological markers that indicate whether a particular piece of information is known to the subject. One such procedure has already been used for a while in combination with the polygraph. The guilty knowledge task (GKT) or concealed information test (CIT) is based on the fact that some knowledge can only be known by the perpetrator of a crime (Lykken, 1959, 1960). Crime-relevant information is embedded in a stream of irrelevant information so that innocent people would not be able to discriminate between relevant and irrelevant items. The assumption is that perpetrators would know the relevant items and thus show a selective arousal response towards the crime-related information, leading to a differential physiological reaction compared to unrelated stimuli. However, even the GKT-polygraph procedure, which is more widely accepted amongst scientists than the CQT, is sensitive to countermeasures (Ben-Shakhar & Dolev, 1996).

The GKT can also be performed with brain-based recording techniques, which might reduce their vulnerability to manipulation. For example, rare and salient stimuli evoke certain electrical signatures in the brain that are measurable at the scalp with EEG. When frequent irrelevant stimuli are alternated with rare probe stimuli (to which the participant is asked to respond), the P300 oddball component is usually larger in response to rare or salient stimuli than to the frequent, irrelevant ones (Donchin & Coles, 1988). In the GKT version of this paradigm, infrequent crime-related stimuli are added as a third stimulus type. For innocent subjects, these stimuli are meaningless and will be treated as one of the frequent irrelevants, not leading to an enhanced P300. Conversely, guilty

subjects with knowledge of the crime will recognise the rare crime-related stimuli, which should lead to an enhanced P300 component in response to these stimuli (Farwell & Donchin, 1991; Rosenfeld, 1988). This method is currently being commercially exploited and claims have been made that it is 100% accurate (www.brainwavescience.com). However, the only peer-reviewed study reporting such high accuracy rates included only six subjects (Farwell & Smith, 2001), and other researchers estimate its accuracy more in the range of 85–95% (Rosenfeld, Soskins, Bosh, & Ryan, 2004), better than the polygraph, but still far below desirable. Moreover, accuracy estimates are based on optimised laboratory conditions, and the only field study employing P300-based measures to date could distinguish liars from truth-tellers only at about chance level (Miyake, Mizutanti, & Yamahura, 1993). Furthermore, like polygraphy-based methods, P300-based paradigms seem vulnerable to countermeasures (Rosenfeld et al., 2004). Finally, the P300 exploits the saliency of a stimulus, which – despite being brain-based – makes it an indirect measure of concealed information.

Both arousal-based and brain-based guilty knowledge tests have the important advantage that they do not necessarily depend on the interviewee's response. Instead, the concealed information is revealed by specific brain responses to the 'guilty' information. Please note, however, that this method still measures concealed information *indirectly* because it requires embedding relevant stimuli in a stream of information and recording *indirect* physiological markers that indicate whether this information is *relevant* to the subject or not. This raises the interesting question whether it might be possible to *directly* read out a person's current thoughts. This would require a method that can directly access the thoughts of a person by decoding them from their current brain activity. Indeed, it has recently emerged that it is possible to measure the contents of specific thoughts of subjects by training a computer to recognise the specific patterns of brain activity associated with each thought (Cox & Savoy, 2003; Haynes & Rees, 2006; Norman, Polyn, Detre, & Haxby, 2006). This can be achieved using the same multivariate pattern recognition techniques that can also be used to improve the efficiency of fMRI lie detection (Davatzikos et al., 2005). However, both techniques are based on recognising patterns of brain activity at different spatial scales. The macroscopic pattern of activity across the entire brain mainly reflects the specific combination of neurocognitive modules involved in processing a task. This can be read out using pattern classification based on the *average* activity in a number of brain regions (i.e., within each brain region activity is averaged across all voxels, see Figure 1b).

The macroscopic decoding method is suitable for detecting whether or not someone is concealing information, but it cannot be used to directly read-out the specific *contents* of a person's thoughts. These contents are encoded in detailed micro-patterns within content-specific brain regions (Figure 1c; Haxby et al., 2001; Haynes & Rees, 2005a,b; Haynes et al., 2007; Kamitani & Tong, 2005; Polyn, Natu, Cohen, & Norman, 2005; Wang, Tanaka, & Tanifuji, 1996). This information is discarded by averaging across all voxels in a region. These micro-patterns reflect a general columnar architecture in which contents are stored in the brain (Mountcastle, 1997). A prototypical and well-studied model of such an architecture is the pattern of orientation columns in primary visual cortex (Hubel & Wiesel, 1969). At first sight, it would appear that the limited resolution of fMRI compared to the fine spatial scale of these columns would preclude measuring column-related signals with fMRI. However, it is possible to exploit subtle fluctuations in the distribution of cells to obtain information about stimulus orientation (Boynton, 2005; Haynes & Rees, 2005a; Kamitani & Tong, 2005). It has been proposed that the whole cortical surface might be arranged in a columnar fashion (Mountcastle, 1997), and examples of more high-level columnar layouts have been presented (Tanaka, 1997, see also Haynes et al., 2007).

Content-based pattern classification allows one in principle to read out arbitrary thoughts a person is having, even potentially without their compliance, or against their will. For example, when people are engaged in thoughts about different types of objects the spatial patterns of activity in object-selective brain regions differ depending on which object a person is thinking about (Haxby et al., 2001). When people are engaged in spontaneous memory retrieval, patterns of cortical activity contain information about the specific category of memories that is recalled (Polyn et al., 2005). In a similar fashion, a pattern classifier might be able to distinguish between different memories a person is trying to conceal. Pattern recognition can even be used to reveal the contents of high-level cognitive representations such as action plans and goals. For example, it is possible to decode from activity patterns in BA 10 in the medial prefrontal cortex which plan a

person is currently holding in mind (Haynes et al., 2007). Neuroimaging experiments have found specific brain activations associated with peoples' attitudes towards brand names (McClure et al., 2004) and towards people of different races (Phelps et al., 2000), with former drug use (Childress et al., 1999), or even sexual preferences (Ponseti et al., 2006). Although these results were obtained with classical univariate group analyses, they clearly reveal the availability of signals that have high potential relevance for classification in forensic questions.

PROBLEMS READING OUT CONCEALED INFORMATION

Although classification of distributed within-area patterns may prove a powerful tool to detect concealed mental contents in individual suspects, it faces a number of similar problems as lie detection, such as the fact that BOLD responses change with age. It also faces some problems of its own, however. Whereas content-based classification using fine-grained voxel patterns retains the highest amount of information present in the fMRI signal, it is less suited to generalise classification across participants. For content-decoding, the mapping between contents and brain patterns needs to be established on a per-subject basis. The reason is that the fine-grained organisation of cortical columns and their sampling with fMRI voxels differs between participants. This is partly due to anatomical differences in the folding patterns of sulci and gyri, and possibly also to the self-organised development of functional patterning in these maps during ontogeny. Hence, a classifier trained on the information coded in voxel-patterns of one participant will most likely perform poorly in other participants. Process-decoding approaches are much less affected by this problem, which explains why lie-detection based on process-decoding is able to generalise across subjects (Davatzikos et al., 2005). However, this comes at the cost of discarding detailed content-related information that is present only at voxel level.

A further problem is that pattern-classification studies have thus far classified brain-activity patterns into only a limited number of categories. In contrast, the number of distinguishable mental states is infinite. Decoding techniques perform pattern *recognition*, not pattern *interpretation*. In order for these algorithms to be able to classify a brain state correctly, they need an extensive training data set to learn to distinguish between a number of mental states. This training requires assigning the correct mental 'labels' to different activation patterns in the training data set, and hence depends on the experimenter knowing the cognitive state of the participant associated with that pattern. Since the number of possible cognitive states is infinite, real 'mind reading' would require an infinite amount of labelled data sets and time to train the classifier. Currently, no known method is able to learn a large set of combinations between brain patterns and thoughts in a short time for a given individual. Thus, a hypothetical 'universal mind-reading machine' will remain science fiction for the foreseeable future. Nonetheless, some applications could arise in the near future that will allow one to distinguish between a finite set of alternative thoughts such as different levels of feelings of familiarity and different degrees of recognition, or different emotional states.

An important question that requires further investigation is whether brain-based measures of deception can also be deliberately manipulated by a trained suspect. The BOLD response on which many fMRI studies rely, is sensitive to variations in respiration, movement artefacts (head motion, swallowing, etc.), and disturbances of the magnetic field. Because they currently cannot be prevented, such factors should be carefully monitored to control for deliberate distortion of results. If distorted data are included in the training dataset and labelled as such, pattern classifiers may learn to distinguish such manipulations. Similarly, the use of cognitive strategies to influence brain-activation patterns may be detected if the training dataset includes trials in which countermeasures are employed. It has been shown, however, that participants can learn to manipulate BOLD activity in specific regions (e.g., Weiskopf et al., 2003). Thus, methods based on BOLD variations in entire brain regions seem potentially vulnerable to such countermeasures. In contrast, fine-grained patterns of brain activity within individual brain regions are certainly much more difficult to deliberately manipulate.

Finally, one could even ask the speculative question whether a person has to be currently thinking about the concealed information in order for a read-out to be possible. It might hypothetically be possible to read out *latent information* that is stored in the memory of the subject but not currently activated. This, however, relies on the availability of a technology that can measure the detailed synaptic connections of the large-scale neural networks in which this latent information is stored. The availability of such techniques for measuring brain *structure* in great detail is not foreseeable for the near future.

Even if they were available, the problem would arise of how to assign *meaning* to specific connection patterns in the neural network. Thus, it seems that the only viable approach at present is to measure information which the person is actively holding in their mind.

PERSPECTIVES

Lie-detection technologies can have an enormous impact on peoples' personal lives, including personal privacy, liberty and job applications. Although some companies (e.g., Brainfingerprinting, NoLieMRI, Cephos) are already actively endorsing or preparing the commercialisation of lie-detection technology, neither of these techniques has been sufficiently validated to warrant its use in daily life. As outlined in the previous sections, their utilisation in real-world, commercial lie-detection application is still problematic for a number of reasons. Carefully constructed field studies are required to validate whether these methods will work reliably also outside the laboratory, be it to detect lies or concealed knowledge. Preferably, these field studies would involve tests in ongoing criminal investigations and thus in the population and scenario they are intended for. By using suspects whose guilt or innocence has not yet been determined a realistic scenario and high motivation is ensured. The classifiers can then be trained as soon as it has independently and unequivocally emerged whether a suspect is guilty or not (see Pollina, Dollins, Senter, Krapohl, & Ryan, 2004, for a similar procedure for polygraph validation). However, care needs to be taken in the design of such field studies, both to ensure the validity of the results and to protect the rights of the participants. For example, strict criteria are necessary to define in which cases it is 'unequivocally' clear whether a suspect is guilty or not, in order to ensure accurate validation. Furthermore, access to the results should be restricted for criminal investigators, in order to prevent the misuse of these preliminary data. Similarly, a suspect's rights should not be affected by their willingness to participate, nor by the final outcomes of the validation process.

There are also other important ethical issues that need to be addressed before promoting commercial applications. It is a fundamental aspect of our mental life that our thoughts are inaccessible to others. Techniques that provide access to peoples' mental states violate their mental privacy (Farah, 2005). Crime-relevant information could potentially be extracted without their consent or even against their will. On the other hand a reliable lie detection technology could also be used to demonstrate a person's innocence, thus making its use an ethical imperative. Thus, lie detection presents a moral dilemma where the rights of the guilty suspect are plotted against the rights of the innocent. The potential future availability of a reliable technology will clearly highlight this dilemma and will require a broad discussion of its ethical consequences (Farah, 2005; Illes & Racine, 2005; Wolpe, Foster, & Langleben, 2005).

CONCLUSION

Recent technological advances in human brain imaging have led to the development of brain-based methods for detecting lies and other forms of concealed information. These promising techniques have clear advantages over conventional lie-detection technology because they directly measure the neurocognitive mechanisms involved in deception and storage of information, rather than relying on peripheral arousal which can be easily manipulated. However, even with these novel techniques, reliable methods for real-world applications are not yet available. A number of fundamental methodological and ethical issues still need to be resolved.

REFERENCES

Abe, N., Suzuki, M., Tsukiura, T., Mori, E., Yamaguchi, K., Itoh, M., et al. (2006). Dissociable roles of prefrontal and anterior cingulate cortices in deception. *Cerebral Cortex, 16*(2), 192–199.

Abe, N., Suzuki, M., Mori, E., Itoh, M., & Fujii, T. (2007). Deceiving others: Distinct neural responses of the prefrontal cortex and amygdala in simple fabrication and deception with social interactions. *Journal of Cognitive Neuroscience, 19*(2), 287–295.

Aron, A. R., Robbins, T. W., & Poldrack, R. A. (2004). Inhibition and the right inferior frontal cortex. *Trends in Cognitive Science, 8*(4), 170–177.

Badre, D., & Wagner, A. D. (2004). Selection, integration, and conflict monitoring; assessing the nature and generality of prefrontal cognitive control mechanisms. *Neuron, 41*(3), 473–487.

Ben-Shakhar, G. (2002). A critical review of the Control Questions Test (CQT). In M. Kleiner (Ed.), *Handbook of polygraph testing*, (pp. 103–126), London: Academic Press.

Ben-Shakhar, G., & Dolev, K. (1996). Psychophysiological detection through the guilty knowledge technique:

Effects of mental countermeasures. *Journal of Applied Psychology, 81*(3), 273–281.

Blasi, G., Goldberg, T. E., Weickert, T., Das, S., Kohn, P., Zoltick, B., et al. (2006). Brain regions underlying response inhibition and interference monitoring and suppression. *European Journal of Neuroscience, 23*(6), 1658–1664.

Botvinick, M., Nystrom, L. E., Fissell, K., Carter, C. S., & Cohen, J. D. (1999). Conflict monitoring versus selection-for-action in anterior cingulate cortex. *Nature, 402*(6758), 179–181.

Botvinick, M. M., Cohen, J. D., & Carter, C. S. (2004). Conflict monitoring and anterior cingulate cortex: An update. *Trends in Cognitive Science, 8*(12), 539–546.

Boynton, G. M. (2005). Imaging orientation selectivity: Decoding conscious perception in V1. *Nature Neuroscience, 8*(5), 541–542.

Bruhn, H., Kleinschmidt, A., Boecker, H., Merboldt, K. D., Hanicke, W., & Frahm, J. (1994). The effect of acetazolamide on regional cerebral blood oxygenation at rest and under stimulation as assessed by MRI. *Journal of Cerebral Blood Flow Metabolism, 14*(5), 742–748.

Buckner, R. L., Snyder, A. Z., Sanders, A. L., Raichle, M. E., & Morris, J. C. (2000). Functional brain imaging of young, nondemented, and demented older adults. *Journal of Cognitive Neuroscience, 12*(Suppl. 2), 24–34.

Burgess, N., Maguire, E. A., & O'Keefe, J. (2002). The human hippocampus and spatial and episodic memory. *Neuron, 35*(4), 625–641.

Childress, A. R., Mozley, P. D., McElgin, W., Fitzgerald, J., Reivich, M., & O'Brien, C. P. (1999). Limbic activation during cue-induced cocaine craving. *American Journal of Psychiatry, 156*(1), 11–18.

Committee to Review the Scientific Evidence on the Polygraph DoBaSSaE (2003). *The polygraph and lie detection*. Washington, DC: National Academies Press.

Courtney, S. M., Petit, L., Haxby, J. V., & Ungerleider, L. G. (1998). The role of prefrontal cortex in working memory: Examining the contents of consciousness. *Philosophical Transactions of the Royal Society of London. Series B, Biological Sciences, 353*(1377), 1819–1828.

Cox, D. D., & Savoy, R. L. (2003). Functional magnetic resonance imaging (fMRI) 'brain reading': Detecting and classifying distributed patterns of fMRI activity in human visual cortex. *Neuroimage, 19*(2 Pt 1), 261–270.

D'Esposito, M., Deouell, L. Y., & Gazzaley, A. (2003). Alterations in the BOLD fMRI signal with ageing and disease: A challenge for neuroimaging. *Nature Reviews in Neuroscience, 4*(11), 863–872.

Davatzikos, C., Ruparel, K., Fan, Y., Shen, D. G., Acharyya, M., Loughead, J. W., et al. (2005). Classifying spatial patterns of brain activity with machine learning methods: Application to lie detection. *Neuroimage, 28*(3), 663–668.

DePaulo, B. M. (1992). Nonverbal behavior and self-presentation. *Psychological Bulletin, 111*(2), 203–243.

DePaulo, B. M., Kashy, D. A., Kirkendol, S. E., Wyer, M. M., & Epstein, J. A. (1996). Lying in everyday life. *Journal of Personal and Social Psychology, 70*(5), 979–995.

Donchin, E., & Coles, M. G. H. (1988). Is the P300 component a manifestation of context updating? *Behavioural Brain Science, 11*(3), 357–427.

Duncan, J., & Owen, A. M. (2000). Common regions of the human frontal lobe recruited by diverse cognitive demands. *Trends in Neuroscience, 23*(10), 475–483.

Ekman, P. (1992). *Telling lies.* New York, NY: W. W. Norton.

Ekman, P., & O'Sullivan, M. (1991). Who can catch a liar? *American Psychology, 46*(9), 913–920.

Elliott, R. (2003). Executive functions and their disorders. *British Medical Bulletin, 65,* 49–59.

Elliott, R., Friston, K. J., & Dolan, R. J. (2000). Dissociable neural responses in human reward systems. *Journal of Neuroscience, 20*(16), 6159–6165.

Farah, M. J. (2005). Neuroethics: The practical and the philosophical. *Trends in Cognitive Science, 9*(1), 34–40.

Farwell, L. A., & Donchin, E. (1991). The truth will out: Interrogative polygraphy ('lie detection') with event-related brain potentials. *Psychophysiology, 28*(5), 531–547.

Farwell, L. A., & Smith, S. S. (2001). Using brain MER-MER testing to detect knowledge despite efforts to conceal. *Journal of Forensic Science, 46*(1), 135–143.

Ford, E. B. (2006). Lie detection: Historical, neuropsychiatric and legal dimensions. *International Journal of Law and Psychiatry, 29*(3), 159–177.

Frith, C. D., Friston, K., Liddle, P. F., & Frackowiak, R. S. (1991). Willed action and the prefrontal cortex in man: A study with PET. *Proceedings of Biological Sciences, 244*(1311), 241–246.

Furedy, J. J., & Heslegrave, R. J. (1991). The forensic use of the polygraph: A psychophysiological analysis of current trends and future prospects. In J. R. Jennings, P. K. Ackles, & M. G. H. Coles (Eds), *Advances in psychophysiology* (pp. 157–189). Greenwich, CT: Jessica Kingsley.

Gamer, M., Rill, H. G., Vossel, G., & Godert, H. W. (2006). Psychophysiological and vocal measures in the detection of guilty knowledge. *International Journal of Psychophysiology, 60*(1), 76–87.

Gamer, M., Bauermann, T., Stoeter, P., & Vossel, G. (2007). Covariations among fMRI, skin conductance, and behavioral data during processing of concealed information. *Human Brain Mapping, 28*(12), 1287–1301.

Ganis, G., Kosslyn, S. M., Stose, S., Thompson, W. L., & Yurgelun-Todd, D. A. (2003). Neural correlates of different types of deception: An fMRI investigation. *Cerebral Cortex, 13*(8), 830–836.

Haxby, J. V., Gobbini, M. I., Furey, M. L., Ishai, A., Schouten, J. L., & Pietrini, P. (2001). Distributed and overlapping representations of faces and objects in ventral temporal cortex. *Science, 293*(5539), 2425–2430.

Haynes, J. D., & Rees, G. (2005a). Predicting the orientation of invisible stimuli from activity in human primary visual cortex. *Nature Neuroscience, 8*(5), 686–691.

Haynes, J. D., & Rees, G. (2005b). Predicting the stream of consciousness from activity in human visual cortex. *Current Biology, 15*(14), 1301–1307.

Haynes, J. D., & Rees, G. (2006). Decoding mental states from brain activity in humans. *Nature Reviews in Neuroscience, 7*(7), 523–534.

Haynes, J. D., Sakai, K., Rees, G., Gilbert, S., Frith, C., & Passingham, R. E. (2007). Reading hidden intentions in the human brain. *Current Biology, 17*(4), 323–328.

Honts, C. R., Raskin, D. C., & Kircher, J. C. (1994). Mental and physical countermeasures reduce the accuracy of polygraph tests. *Journal of Applied Psychology, 79*(2), 252–259.

Horvath, F. (1982). Detecting deception: The promise and the reality of voice stress analysis. *Journal of Forensic Science, 27*(2), 340–351.

Hubel, D. H., & Wiesel, T. N. (1969). Anatomical demonstration of columns in the monkey striate cortex. *Nature, 221*(5182), 747–750.

Illes, J., & Racine, E. (2005). Imaging or imagining? A neuroethics challenge informed by genetics. *American Journal of Bioethics, 5*(2), 5–18.

Kamitani, Y., & Tong, F. (2005). Decoding the visual and subjective contents of the human brain. *Nature Neuroscience, 8*(5), 679–685.

Kozel, F. A., Padgett, T. M., & George, M. S. (2004). A replication study of the neural correlates of deception. *Behavioural Neuroscience, 118*(4), 852–856.

Kozel, F. A., Johnson, K. A., Mu, Q., Grenesko, E. L., Laken, S. J., & George, M. S. (2005). Detecting deception using functional magnetic resonance imaging. *Biological Psychiatry, 58*(8), 605–613.

Langleben, D. D., Schroeder, L., Maldjian, J. A., Gur, R. C., McDonald, S., Ragland, J. D., et al. (2002). Brain activity during simulated deception: An event-related functional magnetic resonance study. *Neuroimage, 15*(3), 727–732.

Langleben, D. D., Loughead, J. W., Bilker, W. B., Ruparel, K., Childress, A. R., Busch, S. I., et al. (2005). Telling truth from lie in individual subjects with fast event-related fMRI. *Human Brain Mapping, 26*(4), 262–272.

Lee, T. M., Liu, H. L., Tan, L. H., Chan, C. C., Mahankali, S., Feng, C. M., et al. (2002). Lie detection by functional magnetic resonance imaging. *Human Brain Mapping, 15*(3), 157–164.

Lee, T. M., Liu, H. L., Chan, C. C., Ng, Y. B., Fox, P. T., & Gao, J. H. (2005). Neural correlates of feigned memory impairment. *Neuroimage, 28*(2), 305–313.

Levin, J. M., Ross, M. H., Mendelson, J. H., Kaufman, M. J., Lange, N., Maas, L. C., et al. (1998). Reduction in BOLD fMRI response to primary visual stimulation following alcohol ingestion. *Psychiatry Research, 82*(3), 135–146.

Lorber, M. F. (2004). Psychophysiology of aggression, psychopathy, and conduct problems: A meta-analysis. *Psychological Bulletin, 130*(4), 531–552.

Lykken, D. T. (1959). The GSR in the detection of guilt. *Journal of Applied Psychology, 43,* 385–388.

Lykken, D. T. (1960). The validity of the guilty knowledge technique: The effect of faking. *Journal of Applied Psychology, 44,* 258–262.

MacDonald, A. W. 3rd, Cohen, J. D., Stenger, V. A., & Carter, C. S. (2000). Dissociating the role of the dorsolateral prefrontal and anterior cingulate cortex in cognitive control. *Science, 288*(5472), 1835–1838.

McClure, S. M., Li, J., Tomlin, D., Cypert, K. S., Montague, L. M., & Montague, P. R. (2004). Neural correlates of behavioral preference for culturally familiar drinks. *Neuron, 44*(2), 379–387.

Miller, E. K., & Cohen, J. D. (2001). An integrative theory of prefrontal cortex function. *Annual Review in Neuroscience, 24*, 167–202.

Milner, B., Squire, L. R., & Kandel, E. R. (1998). Cognitive neuroscience and the study of memory. *Neuron, 20*(3), 445–468.

Miyake, Y., Mizutanti, M., & Yamahura, T. (1993). Event related potentials as an indicator of detecting information in field polygraph examinations. *Polygraph, 22*, 131–149.

Mohamed, F. B., Faro, S. H., Gordon, N. J., Platek, S. M., Ahmad, H., & Williams, J. M. (2006). Brain mapping of deception and truth telling about an ecologically valid situation: Functional MR imaging and polygraph investigation – initial experience. *Radiology, 238*(2), 679–688.

Mountcastle, V. B. (1997). The columnar organization of the neocortex. *Brain, 120* (Pt 4), 701–722.

Norman, K. A., Polyn, S. M., Detre, G. J., & Haxby, J. V. (2006). Beyond mind-reading: Multi-voxel pattern analysis of fMRI data. *Trends in Cognitive Science, 10*(9), 424–430.

Nunez, J. M., Casey, B. J., Egner, T., Hare, T., & Hirsch, J. (2005). Intentional false responding shares neural substrates with response conflict and cognitive control. *Neuroimage, 25*(1), 267–277.

Pavlidis, I., Eberhardt, N. L., & Levine, J. A. (2002). Seeing through the face of deception. *Nature, 415* (6867), 35.

Phan, K. L., Magalhaes, A., Ziemlewicz, T. J., Fitzgerald, D. A., Green, C., & Smith, W. (2005). Neural correlates of telling lies: A functional magnetic resonance imaging study at 4 Tesla. *Academic Radiology, 12*(2), 164–172.

Phelps, E. A., O'Connor, K. J., Cunningham, W. A., Funayama, E. S., Gatenby, J. C., Gore, J. C., et al. (2000). Performance on indirect measures of race evaluation predicts amygdala activation. *Journal of Cognitive Neuroscience, 12*(5), 729–738.

Pineiro, R., Pendlebury, S., Johansen-Berg, H., & Matthews, P. M. (2002). Altered hemodynamic responses in patients after subcortical stroke measured by functional MRI. *Stroke, 33*(1), 103–109.

Pollina, D. A., Dollins, A. B., Senter, S. M., Krapohl, D. J., & Ryan, A. H. (2004). Comparison of polygraph data obtained from individuals involved in mock crimes and actual criminal investigations. *Journal of Applied Psychology, 89*(6), 1099–1105.

Polyn, S. M., Natu, V. S., Cohen, J. D., & Norman, K. A. (2005). Category-specific cortical activity precedes retrieval during memory search. *Science, 310*(5756), 1963–1966.

Ponseti, J., Bosinski, H. A., Wolff, S., Peller, M., Jansen, O., Mehdorn, et al. (2006). A functional endophenotype for sexual orientation in humans. *NeuroImage, 33*(3), 825–833.

Rosenfeld, J. P. (1988). A modified, event-related potential-based guilty knowledge test. *International Journal of Neuroscience, 24*, 157–161.

Rosenfeld, J. P., Soskins, M., Bosh, G., & Ryan, A. (2004). Simple, effective countermeasures to P300-based tests of detection of concealed information. *Psychophysiology, 41*(2), 205–219.

Rother, J., Schellinger, P. D., Gass, A., Siebler, M., Villringer, A., Fiebach, J. B., et al. (2002). Effect of intravenous thrombolysis on MRI parameters and functional outcome in acute stroke <6 hours. *Stroke, 33*(10), 2438–2445.

Rowe, J. B., & Passingham, R. E. (2001). Working memory for location and time: Activity in prefrontal area 46 relates to selection rather than maintenance in memory. *Neuroimage, 14*(1 Pt 1), 77–86.

Ruff, C. C., Woodward, T. S., Laurens, K. R., & Liddle, P. F. (2001). The role of the anterior cingulate cortex in conflict processing: Evidence from reverse stroop interference. *Neuroimage, 14*(5), 1150–1158.

Sell, L. A., Simmons, A., Lemmens, G. M., Williams, S. C., Brammer, M., & Strang, J. (1997). Functional magnetic resonance imaging of the acute effect of intravenous heroin administration on visual activation in long-term heroin addicts: Results from a feasibility study. *Drug and Alcohol Dependence, 49* (1), 55–60.

Smith, E. E., & Jonides, J. (1999). Storage and executive processes in the frontal lobes. *Science, 283*(5408), 1657–1661.

Spence, S. A., Farrow, T. F., Herford, A. E., Wilkinson, I. D., Zheng, Y., Woodruff, P. W. (2001). Behavioural and functional anatomical correlates of deception in humans. *Neuroreport, 12*(13), 2849–2853.

Spence, S. A., Hunter, M. D., Farrow, T. F., Green, R. D., Leung, D. H., Hughes, C. J., et al. (2004). A cognitive neurobiological account of deception: Evidence from functional neuroimaging. *Philosophical Transactions of the Royal Society of London. Series B, Biological Sciences, 359*(1451), 1755–1762.

Starkstein, S. E., & Robinson, R. G. (1997). Mechanism of disinhibition after brain lesions. *Journal of Nervous and Mental Disease, 185*(2), 108–114.

Steinbrook, R. (1992). The polygraph test – A flawed diagnostic method. *New England Journal of Medicine, 327*(2), 122–123.

Tanaka, K. (1997). Mechanisms of visual object recognition: Monkey and human studies. *Current Opinion in Neurobiology, 7*(4), 523–529.

van Veen, V., Cohen, J. D., Botvinick, M. M., Stenger, V. A., & Carter, C. S. (2001). Anterior cingulate cortex, conflict monitoring, and levels of processing. *Neuroimage, 14*(6), 1302–1308.

Verschuere, B., Crombez, G., Koster, E. H., & De Clercq, A. (2007). Antisociality, underarousal and the validity of the Concealed Information Polygraph Test. *Biological Psychology, 74*(3), 309–318.

Vrij, A., & Mann, S. (2001). Telling and detecting lies in a high-stake situation: The case of a convicted murderer. *Applied Cognitive Psychology, 15*, 187–203.

Wang, G., Tanaka, K., & Tanifuji, M. (1996). Optical imaging of functional organization in the monkey inferotemporal cortex. *Science, 272*(5268), 1665–1668.

Weiskopf, N., Veit, R., Erb, M., Mathiak, K., Grodd, W., Goebel, R., et al. (2003). Physiological self-regulation of regional brain activity using real-time functional magnetic resonance imaging (fMRI): Methodology and exemplary data. *Neuroimage, 19*(3), 577–586.

Wolpe, P. R., Foster, K. R., & Langleben, D. D. (2005). Emerging neurotechnologies for lie-detection: Promises and perils. *American Journal of Bioethics, 5*(2), 39–49.

NEUROCASE
2008, 14 (1), 93–121

Sex, aggression and impulse control:
An integrative account

Daniel Strüber,[1,2] Monika Lück,[2] and Gerhard Roth[1,2]

[1]Brain Research Institute, University of Bremen, Bremen, Germany
[2]Hanse-Wissenschaftskolleg – Institute for Advanced Study, Delmenhorst, Germany

There is evidence that the male sex and a personality style characterized by low self-control/high impulsivity and a propensity for negative emotionality increase the risk for impulsive aggressive, antisocial and criminal behavior. This article aims at identifying neurobiological factors underlying this association. It is concluded that the neurobiological correlates of impulsive aggression act through their effects on the ability to modulate impulsive expression more generally, and that sex-related differences in the neurobiological correlates of impulse control and emotion regulation mediate sex differences in direct aggression. A model is proposed that relates impulse control and its neurobiological correlates to sex differences in direct aggression.

Keywords: Aggression; Violence; Psychopathy; Impulsivity; Personality; Sex differences; Molecular genetics; Imaging genomics; Testosterone; Serotonin.

1. INTRODUCTION

Aggressive behavior as such need not be learned; it can be observed even in toddlers at the age of 1–2 years. At this early developmental stage it is also apparent that boys are physically more aggressive than girls (Potegal & Archer, 2004). Across normal development, however, children learn how to deal with aggressive impulses in a socially appropriate manner, and aggression diminishes with age as inhibitory mechanisms mature in adolescence (Tremblay & Nagin, 2005). Nevertheless, sex differences in direct aggression are still present in adults as documented by several meta-analytic reviews (e.g., Bettencourt & Miller, 1996; Eagly & Steffen, 1986; Knight, Fabes, & Higgins, 1996). The sex difference is stronger for physical than for verbal aggression, and men are clearly over-represented in the more violent forms of aggression (Archer, 2004). This effect is also reflected in official crime statistics showing consist-ently higher rates of violent offenses for men than for women, with up to 90% of the most violent crimes like murder and aggravated assault committed by male perpetrators.

One of the most important contemporary theories of crime holds a lack of self-control (i.e., the tendency to pursue short-term, immediate pleasure to the neglect of long-term consequences) to be the sole factor in understanding and predicting criminal activity (Gottfredson & Hirschi, 1990). According to this theory, self-control is established by parental socialization efforts during childhood and then remains stable from 8 to 10 years of age throughout the life span. In interaction with situational opportunities, low self-control facilitates criminal and antisocial behaviors. The strong sex difference in criminal behavior has been traced back to lower levels of self-control in men (Burton, Cullen, Evans, Alardi, & Dunaway, 1998; LaGrange & Silverman, 1999), especially with respect to the subscales

Address correspondence to Daniel Strüber, Hanse-Wissenschaftskolleg, Institute for Advanced Study, Lehmkuhlenbusch 4, 27753 Delmenhorst, Germany (E-mail: dstrueber@h-w-k.de).

DOI: 10.1080/13554790801992743

impulsivity and risk seeking of a personality inventory measuring low self-control (LaGrange & Silverman, 1999).

Similarly, findings from the Dunedin longitudinal study by Moffitt et al. (2001) demonstrated that antisocial behavior is correlated negatively with self-control and positively with negative emotionality (i.e., a propensity for disagreeable affective states including anger and aggression), with males showing more negative emotionality and less self-control/ more impulsivity than the female study participants. The observed sex difference in these personality traits explained 96% of the sex difference in antisocial behavior. Impulsivity and aggression are different psychological constructs and, as different personality traits, are also measured using different psychometric instruments. Yet an overlap between these two constructs is given by the fact that impulsive behavior patterns (e.g., novelty seeking, substance use or recklessness) can also entail a disposition for anger and aggression (Flory et al., 2006). In this context, however, it is important to distinguish between two different types of aggression. While anger and impulsivity are positively correlated with *impulsive aggression* (also labeled 'hostile, reactive, or affective'), there is no correlation with *instrumental aggression* (also labeled 'premeditated, proactive, or predatory') (Ramirez & Andreu, 2006). According to Barratt, Stanford, Dowdy, Liebman, and Kent (1999), impulsive aggression is thoughtless and driven by strong emotions (mostly anger). It occurs as a reaction to perceived threat or provocation and is often followed by remorse. Instrumental aggression, in contrast, is planned, goal-directed (achieving money, status, drugs, etc.), and not necessarily emotionally charged.

In clinical populations, antisocial personality disorder and borderline personality disorder in particular are characterized by a high degree of impulsivity. Correspondingly, expressions of anger and aggression take an impulsive form (Siever, 2005). Impulsive aggression and uncontrollable temper tantrums are also characteristic of intermittent explosive disorder (Coccaro, 2003). A heightened level of instrumental aggression, however, is found exclusively in a special form of antisocial personality known as *psychopathy* (Blair, 2006). Sex-related differences are found in the prevalence rates of these psychiatric disorders. Whereas borderline personality disorder is diagnosed predominantly in women, intermittent explosive disorder, antisocial personality disorder and psychopathy are more frequent in men (American Psychiatric Association,

1994). Characteristic of psychopaths is a lack of fear, empathy, and remorse. Psychopathy is considered a reliable predictor for violence, high rates of recidivism, and poor treatment responsivity (Hare, Clark, Grann, & Thornton, 2000).

Taken together, psychological, criminological, and clinical evidence suggests that the male sex and a personality style characterized by low self-control/ high impulsivity and a propensity for negative emotionality increase the risk for impulsive aggressive, antisocial and criminal behavior. Furthermore, it was shown that sex differences in these personality traits mediate the sex difference in antisocial behavior.

In the following we relate and extend these findings, which are primarily based on psychometric measures, to research on impulsivity and aggression in the neurosciences. We will focus on recent neuroimaging studies in this field as well as genetic, neurophysiological, and hormonal aspects of impulsivity/aggression in normal and abnormal populations. Sex differences will be highlighted throughout, and a model is proposed that relates the neurobiological correlates of impulse control to sex differences in direct aggression.

2. NEUROPSYCHOLOGICAL CORRELATES OF IMPULSIVE AND INSTRUMENTAL AGGRESSION IN ABNORMAL POPULATIONS

2.1 Acquired sociopathy and developmental psychopathy

Patients with bilateral damage to the orbitofrontal cortex (OFC) and the ventromedial prefrontal cortex (VMPFC) often exhibit a disinhibited, socially unsuitable and impulsive behavior. They misinterpret the moods of others, do not comprehend the consequences of their actions, have very little insight into their behavioral problems, exhibit changes with regard to the intensity of emotions, and are impaired in making decisions based on processing emotions (Bechara, 2004; Berlin, Rolls, & Kischka, 2004; Blair & Cipolotti, 2000; Hornak et al., 2003; Mah, Arnold, & Grafman, 2005; Rolls, Hornak, Wade, & McGrath, 1994). However, the symptoms are weaker after circumscribed lesions that are restricted to the OFC without affecting other prefrontal areas (Hornak et al., 2003). In general, the socio-emotional, motivational and moral behavior of such frontal lobe patients is seriously disordered, and accompanied by a personality change that resembles the symptoms of psychopathy. This clinical condition is thus called 'acquired sociopathy' (Damasio, 1994), whereas 'developmental psychopathy' develops during childhood without any apparent brain lesion or marked OFC-related dysfunctions (Blair, Colledge, & Mitchell, 2001; Blair, Colledge, Murray, & Mitchell, 2001).

Despite the neurocognitive parallels, there is an important difference between acquired sociopathy and psychopathy in the predominating kind of aggression that results. While in patients with acquired sociopathy aggressive behavior is exclusively of an impulsive nature (Anderson, Bechara, Damasio, Tranel, & Damasio, 1999; Grafman et al., 1996), psychopathy is characterized by an increased risk of both impulsive and instrumental aggression (Blair, Peschardt, Budhani, Mitchell, & Pine, 2006; Hare, 2001). Blair (2001, 2003) hypothesized that 'cold' instrumental aggression might result from deficits in learning social rules about avoiding antisocial behavior, and considered a dysfunction of the amygdala, which is important for emotional learning, as the primary candidate structure involved in psychopathy. Given the strong interconnections between the amygdala and the OFC (Kringelbach & Rolls, 2004), it was speculated that OFC dysfunction might follow a primary amygdala dysfunction due to a reduction in afferent input from the deficient amygdala to the OFC during the developmental course of psychopathy (Blair, 2004). Two recent imaging studies on adolescents with conduct disorder might be informative on this issue.

An fMRI study by Sterzer, Stadler, Krebs, Kleinschmidt, and Poustka (2005) compared the neural responses to negative emotional pictures in 13 male adolescents with severe conduct disorder (CD) aged 9–15 years and in 14 healthy age-matched control subjects. The results showed an increased activation of the amygdala and the hippocampus in response to the negative pictures compared to neutral pictures. However, the adolescents with CD exhibited a *reduced* responsiveness to negative pictures of the left amygdala and a pronounced deactivation of the right dorsal anterior cingulate cortex (ACC) compared with the control group. Interestingly, differences in OFC activations between CD children and controls were not detected. According to the authors, these results indicate deficits in both the recognition of emotional stimuli and the cognitive control of emotional behavior in CD children, which may be related to their propensity for aggressive behavior. In addition to these *functional* impairments, Sterzer, Stadler, Poustka, and

Kleinschmidt (2007) recently reported *structural* neural deficits in 12 male adolescents with CD by means of magnetic resonance imaging (MRI) in conjunction with voxel-based morphometry. Compared to controls, reduced grey matter volume in CD patients was detected in the left amygdala and the anterior insula bilaterally. Importantly, no effects were found in the ventral or dorsal ACC and OFC (see section 2.4 for a more detailed description of the functional roles of these structures for emotion regulation).

The findings of functional amygdala and dorsal ACC deficits together with reduced anterior insular and amygdala volume in highly aggressive adolescents (Sterzer et al., 2005, 2007) are in line with evidence from fMRI studies showing dysfunctions of these structures in adult psychopaths (Birbaumer et al., 2005; Kiehl et al., 2001; Müller et al., 2003; Veit et al., 2002). Thus, this overlapping pattern of brain abnormalities in CD patients and psychopaths might reflect a corresponding overlap in symptomatology of these disorders. For example, there is evidence for an involvement of the anterior insula and the amygdala in the experience of empathy (Blair, 2005). A lack of empathy is a core symptom of psychopathy and was also described for CD patients by Sterzer et al. (2007). Therefore, the reported deficits in anterior insula and amygdala might underlie the reduced levels of empathy found in CD and psychopathy. From a developmental point of view, it is interesting to note that the adolescents with CD studied by Sterzer et al. (2005, 2007) reveal amygdala deficiencies but lack OFC impairments, whereas adult psychopaths are characterized by both amygdala and OFC dysfunctions (Blair et al., 2006). These findings might be indicative for the above mentioned hypothesis of a primary amygdala deficit in childhood resulting in additional OFC dysfunction by adulthood during the development of psychopathy (Blair, 2004). However, CD is the childhood antecedent of the DSM-IV diagnosis of antisocial personality disorder, which is not identical to the PCL-R based diagnosis of psychopathy. Future studies are warranted to further resolve this issue, ideally by combining structural and functional MRI in children more formally screened for psychopathic tendencies.

2.2 Neurocognitive studies in adult psychopaths

In line with the view of a combined amygdala/OFC dysfunction in adult psychopaths, neuropsychological studies point to impaired functioning of the amygdala and/or the OFC during various types of associative learning and tests of impulsivity (Blair et al., 2006; Budhani, Richell, & Blair, 2006; Lapierre, Braun, & Hodgins, 1995; Mitchell, Colledge, Leonard, & Blair, 2002; Mitchell, Avny, & Blair, 2006a; Mitchell et al., 2006b). However, neuroimaging data on amygdala and OFC activation in psychopaths are inconsistent. Whereas some studies report reduced activation in psychopaths compared to controls using aversive conditioning (Birbaumer et al., 2005; Veit et al., 2002) or affective memory tasks (Kiehl et al., 2001), others found enhanced activations in the amygdala, OFC and dorsolateral prefrontal cortex during emotional learning (Schneider et al., 2000) and the processing of negative emotional pictures (Müller et al., 2003). Using a facial emotion processing task, Deeley et al. (2006) failed to detect any amygdala or OFC activity in criminal psychopaths.

There are several possible reasons for these discrepancies. First, different stimulation paradigms across studies together with generally small sample sizes (*n* < 10 in most of the studies) raise the possibility that some of the reported effects may be sample-specific. Second, the mean total score of the Hare Psychopathy Checklist-Revised (PCL-R; Hare, 1991), which was used consistently to assess psychopathy in the studies cited above, differs considerably across studies (the score of the PCL-R can range from 0 to 40). For example, Birbaumer et al. (2005) reported an overall PCL-R score of 24.9 (range, 15–31), whereas the psychopaths studied by Müller et al. (2003) had a mean score of 36.8 (range, 34–40). Thus, these two samples of psychopaths do not overlap according to their PCL-R score. Third, the PCL-R indexes two distinct, albeit interrelated, facets of psychopathy with Factor 1 reflecting the emotional-interpersonal features (e.g., shallow affect, lack of remorse or empathy), and Factor 2 representing a chronically unstable and antisocial lifestyle (Harpur, Hare, & Hakstian, 1989). It is therefore conceivable that conflicting results on brain activations in psychopaths might partly be due to sample-specific differences in the relation of these two aspects of psychopathy that do not become apparent, when categorizing is based solely on the total PCL-R score.

Overall, the neurocognitive findings on psychopathy suggest that the condition is due to dysfunctions of prefrontal regions, the amygdala, and possibly of additional structures (Blair et al., 2006; Kiehl, 2006; Mitchell et al., 2006b; Soderstrom et al., 2002). In relation to the different types of aggression, impairment of the OFC in particular (as opposed to other sub-structures of the frontal cortex) appears to be associated with impulsive aggression in psychopathy (Blair et al., 2006), while instrumental aggression appears to be linked with dysfunctions of the amygdala and related deficits in passive avoidance learning (Mitchell et al., 2006a). The causes for such dysfunctions of the described brain structures in the development of psychopathy are not yet known. In contrast to acquired sociopathy, there is no defined brain lesion. Recent twin studies in children (Viding, Blair, Moffitt, & Plomin, 2005) and adults (Blonigen, Carlson, Krueger, & Patrick, 2003) provide evidence for a genetic contribution to psychopathy, indicating heritability especially for the affective-interpersonal impairments in psychopathy. Future studies are needed that combine neuroimaging and molecular genetic methodologies in order to reveal potential relations between genetic vulnerability and brain dysfunction in psychopathy.

2.3 Unsuccessful and successful psychopaths

Environmental factors like experiences of violence or abuse in childhood generally lead to enhanced sensitization of the basic threat system and are considered risk factors for increased impulsive aggression (Gollan, Lee, & Coccaro, 2005) and anxiety disorders (Bonne, Grillon, Vythilingam, Neumeister, & Charney, 2004). In contrast, a relation between trauma-based autonomic hyper-responsivity and instrumental aggression in psychopaths seems unlikely, as in this case a *hypo*function of the threat system prevails (Blair et al., 2006; Herpertz et al., 2001; Patrick, Bradley, & Lang, 1993; Veit et al., 2002). A more detailed discussion of this issue is given in section 6.

However, the generalizability of a hypofunction-related conception of psychopathy is questioned by recent studies on distinct subgroups of criminal psychopaths, who have been either convicted for their crimes (*unsuccessful* psychopaths), or have remained undiscovered so far (*successful* psychopaths). In one study, these subgroups were compared with respect to their autonomic stress reactivity and executive functioning (Ishikawa, Raine, Lencz, Bihrle, & LaCasse, 2001). Compared with controls, unsuccessful psychopaths showed reduced cardiovascular stress reactivity, which is in line with previous research. However, in marked contrast to unsuccessful psychopaths, not only did successful psychopaths fail to show hyporeactivity, but they demonstrated even greater autonomic reactivity and stronger executive function than both the unsuccessful psychopaths and the controls. Since both autonomic and executive functional deficits can result from structural damage to the prefrontal cortex (Bechara, Damasio, Tranel, & Damasio, 1997; Damasio, 1994), the hypothesis was tested in another study with the same sample that unsuccessful, but not successful, psychopaths would show prefrontal structural impairments (Yang et al., 2005). Consistent with the hypothesis, there was a 22.3% reduction in prefrontal gray matter volumes specific to the unsuccessful psychopaths compared with control subjects. Prefrontal structural deficits may render unsuccessful psychopaths particularly susceptible to poor decision making, impulsive aggression and unregulated antisocial behavior – thus raising the probability of 'getting caught'. In contrast to unsuccessful psychopaths, successful psychopaths show a relative sparing of prefrontal gray matter that might provide them with normal executive functioning and intact capacities for the control of affective states. Together with high autonomic functioning, this may allow successful psychopaths to react sensitively to environmental cues signaling danger and, therefore, to avoid conviction (Ishikawa et al., 2001; Yang et al., 2005). Although successful psychopaths do not show prefrontal abnormalities, there are several other candidate structures that, when impaired, may give rise to psychopathic features within this group. Most notably, the amygdala awaits further examination in future imaging research on successful and unsuccessful psychopaths.

2.4 Impulsive aggression and the neural circuitry of emotion regulation

Indications of a relationship between impulsive aggression and a reduced frontal lobe volume or other dysfunctions of the prefrontal cortex, especially of the OFC and VMPFC, are found in patients with a variety of psychiatric disorders (Bassarath, 2001; Brower & Price, 2001), including borderline personality disorder (New et al., 2002, 2004a; Soloff, Meltzer, Greer, Constantine, & Kelly, 2000b; Soloff et al., 2003), intermittent explosive disorder (Best, Williams, & Coccaro, 2002; Siever et al., 1999; Woermann et al., 2000), antisocial

personality disorder (Laakso et al., 2002; Raine, Lencz, Bihrle, LaCasse, & Colletti, 2000), in suicidal behavior (Arango, Underwood, Gubbi, & Mann, 1995; Mann, 2003; Oquendo et al., 2003), and in murderers pleading not guilty by reason of insanity (Raine, Buchsbaum, & Lacasse, 1997; Raine et al., 1994). Additionally, it could be shown that patients with intermittent explosive disorder exhibit lower serotonergic innervation in the anterior cingulate cortex (ACC) (Frankle et al., 2005). The serotonin system plays an important role in modulating impulsivity and aggression (see section 4). The ACC with its rostral 'affective' and dorsal 'cognitive' divisions is important for the automatic and effortful regulation of affective states and emotional behavior (Phillips, Drevets, Rauch, & Lane, 2003a), as further outlined below.

Sex differences in relation to prefrontal dysfunction and impulsive aggression are clearly understudied in clinical and forensic populations. To the best of our knowledge, there is only one study examining patients with borderline personality disorder (eight women, five men) separated by sex (Leyton et al., 2001). Impulsive behavior was not investigated in direct relation to aggression, but assessed by using a laboratory measure of behavioral disinhibition. Neurotransmission of serotonin in the brain was measured by means of positron emission tomography (PET). Compared to healthy controls, female patients had lower serotonin synthesis capacity in the right superior temporal gyrus and right dorsal ACC, while for male patients reductions were observed in the medial frontal cortex, rostral ACC, and orbitofrontal cortex. This sex difference might be related to differences in the behavioral phenotype between men and women, because two of the five males met the criteria for antisocial personality disorder. In addition, all of the men but only half of the women had a history of past substance use disorder, whereas more women than men had a history of binge eating. However, negative correlations with impulsivity scores were identified for both men and women in similar brain regions, including the medial frontal gyrus and the rostral and dorsal ACC bilaterally.

Collectively, the above-mentioned studies provide strong evidence that structural or functional prefrontal impairments are associated with a heightened risk of impulsive aggression and violent behavior. However, prefrontal areas do not act in isolation. The prefrontal cortex, in particular the OFC, is massively interconnected with limbic brain areas such as the rostral ACC and the amygdala

(Kringelbach & Rolls, 2004). Accordingly, it is more likely that the relative balance of activity between the prefrontal cortex and limbic structures might be disturbed when it comes to impulsive aggression and violence, rather than impairments in one or the other structure alone (Bufkin & Luttrell, 2005). There are, however, relatively few studies examining amygdala–OFC interrelations in impulsive aggression and violence specifically. In a PET study, Raine et al. (1998) assessed the ratio of prefrontal-to-subcortical brain functioning in a group of murderers compared to controls. Reduced prefrontal and increased subcortical (including the amygdala) metabolism was found specifically in those murderers who committed their violent acts in an unplanned and impulsive manner. Similarly, patients with intermittent explosive disorder showed an exaggerated amygdala reactivity and diminished OFC activation in response to faces expressing anger (Coccaro, McCloskey, Fitzgerald, & Phan, 2007b). These latter fMRI findings by Coccaro et al. (2007b) provide evidence for amygdala–OFC dysfunction in response to a social threat signal in individuals with impulsive aggression. Dougherty et al. (2004) demonstrated an aberrant functional relationship between the VMPFC and the amygdala during anger-inducing script imagery in depressed patients with anger attacks. Furthermore, patients with borderline personality disorder exhibited simultaneous amygdala and OFC volume loss (Tebartz van Elst et al., 2003). Taken together, these imaging findings substantiate a link between a dysfunctional frontal-limbic circuitry and hyperarousal-related disorders characterized by impulsive aggression and emotional dysregulation.

The amygdala and OFC are key structures in the ventral system of emotion perception as described by Phillips et al. (2003a). According to the authors, the ventral system, including the amygdala, insula, ventral striatum, rostral ACC, and ventral regions of the prefrontal cortex (including the OFC), is important for the rapid appraisal of emotional material, the production of affective states, and the automatic regulation of autonomic responses to emotional stimuli. Furthermore, these authors suggested that specific abnormalities within the ventral system are associated with disturbances in emotional regulation and behavior that are characteristic of different psychiatric disorders, such as schizophrenia and bipolar and major depressive disorder (Phillips, Drevets, Rauch, & Lane, 2003b). Interestingly, heightened levels of anger and aggression, which is commonly found in patients with these disorders, could be related to abnormalities in frontal-limbic structures of the ventral system (Dougherty et al., 2004; Hoptman et al., 2002).

In line with the presented evidence linking emotional dyscontrol, impulsive aggression, and frontal-limbic abnormalities, it has been proposed that dysfunctions of the network responsible for the regulation of emotion, including the OFC, ACC, and amygdala, can lead to impulsive and violent behavior in at-risk populations (Davidson, Putnam, & Larson, 2000). According to this proposal, the ACC and OFC normally regulate the intensity of negative emotion via an inhibitory connection to the amygdala (see Figure 1, left column, 'Neuronal structures'; arrow indicates inhibitory input from ACC/OFC to the amygdala which generates fear or anger in response to situational aspects signaling threat).

2.5 Summary and conclusion

Prefrontal disruption consistently occurs across anatomical and functional studies with a wide range of clinical and forensic populations, which are characterized by emotional dyscontrol and impulsivity/aggression, suggesting that prefrontal dysfunction may predispose to impulsive aggression and violence. However, the prefrontal cortex, in particular the OFC, is just one structure within the ventral system of emotion regulation, including limbic areas such as the rostral ACC and the amygdala as well. Thus, studies examining relative activations in both regulatory (ACC, OFC) and emotion-generating (amygdala) structures at the same time in impulsive aggressive populations are of special importance for a better understanding of impulsive aggression in terms of a regulating balance within the neural circuitry of emotion regulation. Whereas impulsive aggression is related to a variety of psychiatric disorders, heightened levels of both impulsive and instrumental aggression are found exclusively in individuals with psychopathy. In general, neurocognitive findings in adult psychopaths point to dysfunctions of both the amygdala and OFC. Developmentally, psychopathy may be due to a primary amygdala deficit as indicated by a hypofunction of the basic threat system and the disruption of specific forms of emotional learning. Thus, affected children are prevented from learning to avoid antisocial behavior, which, in turn, raises the risk for engaging in instrumental aggression. In conclusion, the current evidence suggests OFC hypofunction as a common risk factor for impulsive aggression in many disorders, including psychopathy. In combination with a hyperresponsiveness of the amygdala to emotional stimuli, the risk of uncontrollable aggressive outbursts further increases, as seen for example in intermittent explosive disorder. Increased instrumental aggression is primarily associated with a hypofunction of the amygdala.

3. NEURAL CORRELATES OF IMPULSIVITY, ANGER, AND AGGRESSION IN NORMAL VOLUNTEERS AND ITS RELATION TO CLINICAL AND PSYCHOSOCIAL DATA

3.1 Impulsivity

As mentioned in the introduction, impulsivity is highly correlated with impulsive aggression (Ramirez & Andreu, 2006). Furthermore, Flory et al. (2006) demonstrated that dispositional impulsivity was consistently represented by the same features (labeled thrill seeking, nonplanning and disinhibited behavior) in a normal and a personality-disordered sample (e.g., borderline and antisocial personality disorder). Interestingly, these findings indicate that personality disorders characterized by failures of impulse control represent the maladaptive extreme of a continuum of impulsive-aggressive personality traits (Flory et al., 2006). In light of the evidence suggesting frontal-limbic dysfunction as a brain correlate of emotional dysregulation and impulsive aggression in personality-disordered patients (see section 2.4), the question arises whether a similar relationship between the functioning of the neural circuitry of emotion regulation and impulsive personality is seen in normal volunteers.

Related to this issue, Brown et al. (2006) tested 58 volunteers in an fMRI-study by employing well-established paradigms that reliably evoke either prefrontal inhibitory responses (*prefrontal inhibitory control paradigm;* go/no-go paradigm) or arousal-related amygdala reactivity (*amygdala reactivity paradigm;* face-processing task). Brain activations were correlated with dispositional impulsivity. The results showed a negative correlation between impulsivity and activations of the lateral OFC and the dorsal amygdala, whereas impulsivity correlated positively with activations of the bilateral ventral amygdala, parahippocampal gyrus, dorsal ACC, and bilateral caudate. Although these results point to an involvement of frontal-limbic areas in the normal range of

dispositional impulsivity, the pattern of activations differ from that typically seen in patients with impulse control problems. In particular, the differential association of the dorsal and ventral amygdala with low and high impulsivity, respectively, is not as clear as the relation between amygdala hyperresponsiveness and impulsivity found in impulsive patients (Coccaro et al., 2007b). With regard to the dorsal ACC, Brown et al. (2006) described an increased activation for more impulsive individuals during the go/no-go paradigm, whereas CD patients, also characterized by heightened impulsivity, showed a marked decrease in the dorsal ACC when viewing negative pictures (Sterzer et al., 2005). Possibly, activation of the dorsal ACC in normal impulsive individuals operates as a compensatory mechanism, which is absent in clinical populations. The use of different paradigms might also explain the inconsistent findings. However, the negative correlation between dispositional impulsivity and activation of the lateral OFC in the study by Brown et al. (2006) during the face-processing task is compatible with the clinical data suggesting OFC hypofunction in impulsive patients, as reviewed in section 2.4. Conversely, Horn et al. (2003) found greater activation of the right inferior frontal gyrus and posterior OFC in more impulsive individuals when employing the go/no-go paradigm, indicating that greater engagement of these regions was needed to maintain behavioral inhibition in more impulsive individuals. This points towards a compensatory role for the OFC in normal impulsivity, as was suggested by Brown et al. (2006) with regard to increased activity of the dorsal ACC during the go/no-go task. Taken together, studies on impulsivity-related brain activations in normal volunteers suggest a role for the posterior OFC, the dorsal ACC, and the amygdala. However, the data are limited, and further research is necessary in order to characterize brain correlates of impulsive traits as varying on a continuum from normal to abnormal, as was demonstrated on the psychometric level (Flory et al., 2006).

With regard to sex differences in emotion regulation, there is neuroanatomical evidence for a higher ratio of orbital gray to amygdala volume in women compared to men (Gur, Gunning-Dixon, Bilker, & Gur, 2002). That is, both sexes had identical amygdala volumes but women had larger OFCs than men. In line with this anatomical finding, Meyer-Lindenberg et al. (2006) demonstrated in an fMRI study a decreased functional connectivity between OFC and amygdala for men compared to women

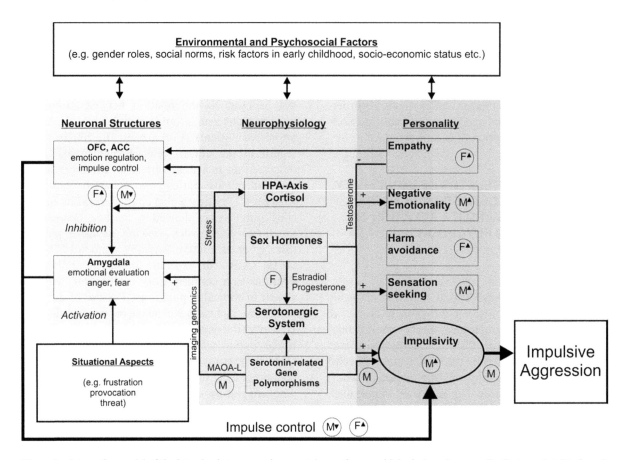

Figure 1. Integrative model of the interplay between environmental, genetic, neurobiological, and personality factors related to impulsive aggression, with a special emphasis on sex differences. The links are explicated in the text.

Note: OFC, orbitofrontal cortex; ACC, anterior cingulate cortex; HPA-axis, hypothalamic–pituitary–adrenal axis; M, present in males; F, present in females; (M↑), more pronounced in males compared to females; (M↓), less pronounced in males compared to females. (F↑), more pronounced in females compared to males.

during a face-matching task, indicating that the OFC-based regulation of the amygdala is intrinsically weaker in men. These sex differences are depicted in Figure 1 (left column; arrow from OFC to amygdala, different strength of inhibition for males and females is indicated). It is suggested that these results provide a neuronal basis for sex differences in emotion regulation and impulse control that are reflected in the male preponderance for impulsivity and aggression (see Figure 1, arrow 'Impulse control' links OFC, amygdala and impulsivity, with impulsive aggression as the behavioral output).

3.2 Anger

Anger is a common response to provocation and is positively related to impulsivity and impulsive aggression (Ramirez & Andreu, 2006). One might

therefore expect the involvement of frontal-limbic structures implicated in emotion regulation.

Most neuroimaging studies examining the neural basis of anger used either angry facial expressions as stimuli or employed an anger induction paradigm in which subjects recalled intense autobiographic episodes involving anger in comparison to neutral induction. Three of these studies used angry facial expressions and found increased activity in the left OFC (Sprengelmeyer, Rausch, Eysel, & Przuntek, 1998), and in the right OFC and rostral ACC (Blair, Morris, Frith, Perrett, & Dolan, 1999; Phillips et al., 1999). Anger induction also activated the (left) OFC/VMPFC in most studies (Dougherty, Bjork, Marsh, & Moeller, 1999; Dougherty et al., 2004; Kimbrell et al., 1999), although Damasio et al. (2000) reported a *decrease* of OFC activity during anger induction, together with activations of the midbrain and pons (also not noted to be active in other anger studies). Thus, with one exception, anger was consistently associated with increased activation of the OFC across all studies. The exceptional decrease of OFC activity in the Damasio et al. (2000) study might be related to specific details of content or intensity of the recalled episodes, which are difficult to fully control in an emotion induction paradigm based on the subjective recall of personal episodes. Possibly, asking the participants to re-experience and re-enact intense episodes of anger may have also involved aggressive acts. The switch from feeling angry to behaving aggressively might be corresponded by a marked decrease of OFC activity. In line with this interpretation, a functional decrease in the activity of the medial OFC during imaginal aggressive behavior had been found in a PET study (Pietrini, Guazzelli, Basso, Jaffe, & Grafman, 2000), where the participants had to imagine that they were seriously attacking two men with all their strength, who were assaulting the subject's mother in an elevator. Thus, the OFC, when activated by experiencing anger, normally acts to suppress overt aggressive behavior. In consequence, patients exhibiting excessive aggressive behavior might do so, because they cannot mobilize the OFC in response to experienced anger. In support of this suggestion, Coccaro et al. (2007b) found in patients with intermittent explosive disorder a hypofunction of the OFC in response to angry faces. Similarly, depressed patients with increased levels of anger and aggression failed to show VMPFC activation during anger induction (Dougherty et al., 2004).

As regards the amygdala, none of the above-mentioned studies on anger found any amygdala activation in healthy subjects. However, more recent imaging data provide evidence for an amygdala responsiveness to angry faces (Adams, Gordon, Baird, Ambady, & Kleck, 2003; Fitzgerald, Angstadt, Jelsone, Nathan, & Phan, 2006; Strauss et al., 2005; Whalen et al., 2001). Nomura et al. (2004) demonstrated an inverse relationship between the response of the amygdala and the lateral OFC during the cognitive evaluation of facial expressions primed by masked angry faces in an fMRI study. This finding indicates amygdala reactivity to angry faces even during subliminal perception and further underlines the involvement of a frontal-limbic circuitry of emotion regulation in response to facial expressions signaling threat.

One study examined sex differences in OFC and amygdala engagement during the evaluation of threat induced by angry faces (McClure et al., 2004).

In addition, activations associated with neutral, happy, and fearful faces were examined with fixation as baseline. Women showed greater OFC and amygdala activation to angry versus neutral faces than men. Further analyses revealed frontal-limbic activations also in response to neutral and fearful faces in men, when compared to baseline, whereas frontal-limbic activations in women were specific to angry faces. These neuroimaging findings indicating a specific sensitivity to threat in women might relate to psychometric findings showing higher scores for women than men in the personality trait harm avoidance (i.e., the dislike of dangerous activities and unwillingness to expose oneself to attack and injury are congruent with), as outlined in section 3.4.

3.3 Aggression

Only recently, researchers began to design dynamic and social interactive scenarios that allow the examination of the neural correlates of normal aggressive behavior under semi-naturalistic conditions. Brain–environment interactions were demonstrated using fMRI during a violent, interactive video game ('first-person shooter game') in 13 male experienced gamers (Mathiak & Weber, 2006). The results showed *increased* activity during violent episodes in the dorsal (cognitive) part of the ACC, while the rostral (affective) part of the ACC, together with the amygdala and the OFC, were *deactivated*. According to the authors, these results suggest that affective processing is actively suppressed during virtual violence in favor of the cognitive operation. This pattern of results is incongruent with findings of increased OFC and rostral ACC activations during anger in normal volunteers (section 3.2), and with patient data showing decreased activity in the dorsal ACC (Sterzer et al., 2005) and exaggerated amygdala reactivity (Coccaro et al., 2007b) in response to negative emotional material. Arguably, the results of Mathiak and Weber are not related to impulsive aggression, rather than reflecting more instrumental aspects of aggression and violence. It seems plausible to assume that experienced players of first-person shooter video games do not act impulsively, when virtually killing opponents, but are motivated to reach a goal, i.e., winning the game. To do so, players have to navigate through a complex virtual environment, react precisely in a violent situation, and virtually kill opponents. Conceivably, this kind of violence is primarily regulated by effortful control mechanisms (as reflected by dorsal ACC activity) resulting in a functional 'shut-down' of brain areas involved in the affective processing (amygdala, rostral ACC, OFC). Interestingly, a similar pattern of deactivations has been described for psychopaths (see sections 2.1 and 2.2) further supporting the notion that violence in first-person shooter games is instrumentally motivated and might be accompanied by a (temporary) reduction of empathy.

The difference between empathic and violent behavior was analyzed in a further fMRI study (King, Blair, Mitchell, Dolan, & Burgess, 2006) by varying the social appropriateness of aggressive and empathic behavior using a customized video game. Twelve healthy participants (six male) were offered virtual environments, in which they were supposed to either shoot an aggressive humanoid assailant or bandage a wounded conspecific (appropriate behavior). Inappropriate behavior demanded the converse combination, i.e., shooting the injured conspecific and healing the humanoid attacker. Surprisingly, identical activations were observed in a circuit including the VMPFC and the amygdala for the appropriate situations, irrespective of whether the behavior was violent or compassionate. This finding suggests that context-appropriate behavior is guided by a common neural system including the VMPFC and amygdala. These data also support suggestions that dysfunction in this system underlies the presentation of inappropriate social behavior as seen in acquired sociopathy and psychopathy (see section 2.1).

In an interactive fMRI study, Lotze, Veit, Anders, and Birbaumer (2007) examined the role of the prefrontal cortex in the regulation of reactive aggression. In a competitive reaction time task, 16 male volunteers were provoked by a virtual opponent who was allowed to administer aversive pain stimuli to the subjects when he won the challenge. However, when the opponent lost, there was the opportunity for revenge, whereby the participants could determine the intensity of the penalty and watch the opponent suffer on video tape. Subjects were classified in terms of their psychopathic traits (in the normal range) into high and low callous participants. Increased activity was observed in the medial prefrontal cortex (mPFC) during retaliation, and in the dorsal mPFC, when subjects had to select the revenge intensity. The dorsal mPFC activation also correlated positively with the intensity of the selected stimulus strength when taking revenge, but not when watching the opponent suffering. The OFC/VMPFC was active during watching the opponent suffering; however, this effect was stronger for low callous (i.e., more empathic) participants. The amygdala was also found to be active throughout the different phases of the experiment, and the activity of the right amygdala during watching the opponent correlated positively with the intensity of pain of the opponent. According to the authors, the results are indicative for a participation of the dorsal mPFC in cognitive aspects of social interaction, while the OFC/VMPFC is involved in affective processes associated with compassion to the suffering opponent. With regard to activations of OFC/VMPFC and amygdala during viewing the opponent suffer, the findings resemble those of King et al. (2006) in that punishing an opponent for his unjustified aggression might be experienced as an appropriate social behavior, irrespective of being aggressive. Interestingly, Lotze et al. (2007) reported stronger OFC/VMPFC activations for more empathic participants and a positive correlation between amygdala activity and pain of the opponent. These findings point towards a higher responsiveness of the OFC/VMPFC to distress cues signaled by the amygdala in more empathic individuals and may relate to a neural basis of empathy deficits as seen in psychopaths. In addition, Lotze et al. (2007) found stronger activation of OFC/VMPFC in more empathic participants during adjustment of the retaliation intensity, suggesting a more effective control of anger in response to provocation and, as a consequence, lower levels of aggressive behavior compared to low empathic participants. Together, these observations by Lotze et al. (2007) provide neuropsychological evidence for a role of empathy in the regulation of aggressive behavior.

Using a similar competitive reaction time task as Lotze et al. (2007), Krämer and co-workers (Krämer, Jansma, Tempelmann, & Münte, 2007) found activations in the anterior insula and the rostral and dorsal ACC, when the participants had to select the intensity of the punishment following high compared to low provocation. These activations probably reflect negative emotions (anger or disgust) and higher arousal elicited under high provocation. Activations in the dorsal striatum and dorsal ACC were observed when subjects selected high punishments in response to high provocation, indicating reward expectations (dorsal striatum) and higher cognitive efforts (dorsal ACC) during the decision to retaliate. Reward processing during the aggressive act (punishment) itself in win trials was signaled by activation of the ventral striatum. In contrast to the findings of Lotze et al. (2007), no aggression-related activity in the VMPFC or the dorsal mPFC was found. As regards the dorsal mPFC, this inconsistency might be due to differences in the experimental design. Whereas Krämer et al. (2007) introduced two opponents to the participants – one highly and one less provoking – Lotze et al. (2007) introduced one opponent who administered pain stimuli of increasing intensity over the course of the experiment. Conceivably, being confronted with steadily increasing provocation independent of one's own reaction might induce 'mentalizing', i.e., figuring out what others might intend during social interactions. Since the dorsal mPFC is one key structure in mentalizing (e.g., Kampe, Frith, & Frith, 2003), the engagement of this region during selection of the retaliation intensity in the Lotze et al. (2007) study might partly reflect mentalizing-related activity. In contrast, the opponents' behavior in the study of Krämer et al. (2007) was highly predictable, thus reducing the need for mentalizing and dorsal mPFC engagement.

So far, sex differences have not been considered in neuroimaging studies on aggression in normal volunteers, although several fMRI studies demonstrated sex differences in the processing of various emotions (Hamann & Canli, 2004). However, a recent study by Krämer, Büttner, Roth, & Münte (in press) examining the influence of trait aggressiveness on event-related potentials (ERPs) in the competitive reaction time task described above, also probed for putative sex differences. The behavioral results of this study confirmed the well-known observation of higher trait aggressiveness in men than in women (see section 3.4). Interestingly, high trait-aggressive men also showed more actual aggressive behavior in the competitive reaction time task, whereas trait aggressiveness in women was not predictive for aggressive behavior. However, this sex difference in the relation between trait aggressiveness and behavior was not reflected in the ERP data.

3.4 Psychological evidence for sex differences in aggression-related personality traits

As outlined in the introduction, psychosocial studies substantiate that boys/men are physically more aggressive than girls/women (Eagly & Steffen, 1986; Potegal & Archer, 2004). Provocation diminishes the magnitude of this sex difference (Bettencourt & Miller, 1996), indicating that emotional arousal may be an important variable for sex differences in

aggression. There is, indeed, evidence from a meta-analysis that men and women differ in their ability to regulate their emotional arousal and that this difference may underlie the observed sex difference in direct aggression (Knight, Guthrie, Page, & Fabes, 2002). Knight et al. (2002) found that sex differences in aggression were small at both very low and very high levels of arousal, because at low levels both sexes were able to regulate their emotions and at very high levels the regulation capacities were equally disrupted in men and women. However, at intermediate levels of arousal, the weaker ability to control emotions in men led to higher sex differences in aggressive behavior. Although these findings might suggest that men experience more anger, there is no evidence of a sex difference in the frequency and intensity of anger (Archer, 2004). However, women appear to express their anger only after reaching a significantly higher level of excitation, which results in a more expressive experience of losing self-control, whereas men typically view their aggression as seizing control of the situation and exerting dominance (Campbell & Muncer, 1994).

In a current study, it was tested whether this sex difference in social representations of aggression can be explained by differences in inhibitory control (Driscoll, Zinkivskay, Evans, & Campbell, 2006). The results showed that self-reported behavioral constraint significantly mediates the relationship between sex and social representation of aggression. The strongest contribution was found for harm avoidance (i.e., the dislike of dangerous activities and unwillingness to expose oneself to attack and injury), followed by the subscale control versus impulsivity, as measured with the Multidimensional Personality Questionnaire (MPQ). More specifically, women were markedly higher on the MPQ main variable constraint, and this female advantage was reflected in their higher scores on the subscales harm avoidance and control vs. impulsivity. With respect to the main variable negative emotionality, women scored higher on the subscale stress reaction, whereas a male advantage was found for the subscale aggression. In line with these findings are recent results from Verona and Kilmer (2007), showing in a laboratory aggression paradigm that men exposed to high stress exhibited increases in aggression relative to those under low stress, whereas women under high stress responded with less aggression than women under low stress. Moreover, there is longitudinal evidence that the sex difference in the MPQ variables negative emotionality and impulsivity (both in the male direction) explained 96% of the sex difference in antisocial behavior (Moffitt et al., 2001). Taken together, there is evidence to suggest that higher levels of constraint in women, as reflected by their lower impulsivity and higher harm avoidance scores, mediate the sex difference in direct aggression.

In this context, it has been proposed that behavioral inhibition rests on an infrastructure of fear and that women's greater fear of injury plays a stronger role for sex differences in aggression than women's higher behavioral restraint (Campbell, 2006). Furthermore, empathy is negatively related to aggression and antisocial behavior, whereas a positive association exists with higher levels of fear and effortful control (Campbell, 2006). Together with the documented sex difference in empathy favoring women (Baron-Cohen, Knickmeyer, & Belmonte, 2005), these findings suggest a modulating role for empathy with regard to sex differences in emotion regulation and aggression. In contrast, sensation seeking (i.e., the search for exciting and adventurous activities) is positively associated with impulsivity and shows a strong sex difference in the male direction (Roberti, 2004). The preponderance of males in impulsivity and risk seeking has been related to their higher engagement in delinquency and violence (LaGrange & Silverman, 1999). The aggression-related personality traits, as described in this section, are listed in Figure 1 (right column 'Personality'; direction of sex difference is indicated). Of note, there are indications for greater OFC activation in more empathic (low callous) individuals during reactive aggression (Lotze et al., 2007) as reported in section 3.3. This relationship between empathy and brain function is indicated in Figure 1 (link from empathy in the right column to OFC in the left column).

3.5 Summary and conclusion

Recent neuroimaging studies on impulsivity, anger, reactive aggression, and virtual violence during naturalistic scenarios in normal volunteers demonstrate consistently the involvement of brain structures implicated in the neural circuitry of emotion regulation, including prefrontal areas (primarily OFC/VMPFC), medial frontal cortex (rostral and dorsal ACC) and limbic structures (amygdala) in healthy volunteers. However, the combination of these structures engaged during a specific task or scenario as well as the direction of the effects, i.e., whether regions are activated or deactivated, varies according to the specific construct under study. As regards impulsivity, the existing data on the association between dispositional impulsivity and brain activations points to an involvement of the lateral OFC, the dorsal ACC, and the amygdala. The induction of anger activates the OFC in healthy persons and probably serves to suppress overt aggression, as indicated by OFC hypofunction in patients exhibiting excessive aggressive behavior. Studies examining OFC-amygdala interaction in response to anger report an inverse relationship in normal controls, which is disturbed in patients with anger attacks and impulsive aggression. Recent studies on the neural underpinnings of aggression are more diverse with respect to the paradigms employed. Brain correlates of playing first-person shooter games point to more instrumental aspects of aggression, as indicated by activation increases in cognition-related areas and deactivation of affective regions. In contrast, interactive fMRI studies on reactive aggression demonstrate the engagement of brain regions related to both cognitive and affective processing. So far, interactive fMRI studies on aggression have not been used in patients, leaving a direct comparison with normal controls for future research.

There are indications from neuroimaging studies for sex differences in the structural and functional relationship between the amygdala and OFC suggesting weaker emotion regulation capacities for men compared to women. Similarly, psychological evidence points toward a mediating role of sex differences in emotion regulation for sex differences in direct aggression. In addition, there are indications that men score higher on personality dimensions that are positively correlated with aggression and negatively with behavioral inhibition (impulsivity, negative emotionality, sensation seeking), whereas women score higher on dimensions that are negatively associated with aggression and positively with behavioral inhibition (control, harm avoidance, empathy). Thus, there is neuroimaging and psychological evidence consistent with the hypothesis that sex differences in behavioral inhibition/impulsivity mediate sex differences in direct aggression.

4. NEUROPHYSIOLOGICAL AND GENETIC ASPECTS OF IMPULSIVITY AND AGGRESSION

4.1 Serotonin (5-HT)

A central serotonin deficit has long been associated with increased impulsivity and impulsive aggression

(Lee & Coccaro, 2001; Lesch & Merschdorf, 2000; Virkkunen et al., 1994). Such a relationship has been described for various clinical populations by a variety of different methods (Mann, 2003; New et al., 1998, 2002, 2004b; Placidi et al., 2001; Soloff, Lynch, & Moss, 2000a; Soloff et al., 2000b, 2003; Stanley et al., 2000).

This association can be considered well-documented for healthy subjects as well (Bjork, Dougherty, Moeller, & Swann, 2000; Bond, Wingrove, & Critchlow, 2001; Cleare & Bond, 1997; Dougherty et al., 1999; Manuck et al., 1998; Manuck, Flory, Ferrell, Mann, & Muldoon, 2000; Manuck, Flory, Muldoon, & Ferrell, 2002; Marsh, Dougherty, Moeller, Swann, & Spiga, 2002; Netter, Hennig, & Roed, 1996; Wingrove, Bond, Cleare, & Sherwood, 1999a; Wingrove, Bond, Cleare, & Sherwood, 1999b). Furthermore, there are a number of studies that describe a reduced serotonergic modulation of frontal brain areas in the context of impulsivity/aggression (Arango et al., 1995; Frankle et al., 2005; New et al., 2002, 2004a; Rubia et al., 2005; Siever et al., 1999; Soloff et al., 2003). These studies provide neuro-functional evidence for the suggested link between OFC/ACC, serotonin, and inhibitory control. More specifically, they suggest that OFC and adjacent regions exert an inhibitory effect on impulsive aggression through a serotonergic mechanism.

Sex differences in the linkage between diminished central serotonergic function and impulsivity/aggression have been documented for personality disordered patients with impulsive aggression (New et al., 2004b; Soloff, Kelly, Strotmeyer, Malone, & Mann, 2003), violent criminals (Moffitt et al., 1998), and healthy subjects (Cleare & Bond, 1997; Manuck et al., 1998; Møller et al., 1996). These studies consistently found significant relationships for men only, but two studies using dietary depletion of tryptophan (an amino acid needed for the synthesis of central 5-HT) in healthy female samples reported that decreased serotonergic neurotransmission increases aggression in women, too (Bond et al., 2001; Marsh et al., 2002). If there is a sex difference in the relation between central serotonergic function and impulsive aggression, this might be associated with a more general sex difference in the functioning of the serotonergic system. This is indicated by findings showing that the mean rate of serotonin synthesis in the brains of normal males is 52% higher than in normal females (Nishizawa et al., 1997). In addition, sex differences in the binding capacity of serotonin receptors were found in several brain regions including regulatory structures such as the ACC and OFC (Biver et al., 1996; Parsey et al., 2002). Parsey et al. (2002) reported a negative correlation between binding potential and self-reported aggression for both sexes, indicating a possible anti-aggressive effect mediated by certain serotonin receptors (5-HT$_{1A}$). A relation between blunted 5-HT$_{1A}$ receptor function and increased impulsivity/aggression has also been demonstrated in pharmacological challenge studies (Cleare & Bond, 2000; Coccaro, Gabriel, & Siever, 1990; Minzenberg et al., 2006).

Sex differences in serotonergic functioning are likely related to an interaction with ovarian steroids. Estradiol and progesterone are known to affect the density of certain serotonin receptor sites and other aspects of the serotonergic system in women (Rubinow, Schmidt, & Roca, 1998; Wihlbäck et al., 2004). Furthermore, there are sex differences in the effect of serotonin-related gene polymorphisms on the central serotonergic function (Williams et al., 2003). Given its role for impulse control and the documented sex differences, the serotonin system qualifies as an important mediator for sex differences in aggression. The modulation of the serotonin system by female sex hormones (estradiol, progesterone) and the link between the serotonin system and prefrontal structures (ACC, OFC) is illustrated in Figure 1 (middle column, arrow pointing from sex hormones to the serotonergic system and from the serotonergic system to the inhibitory connection between OFC/ACC and amygdala, respectively).

4.2 Genetic aspects of the serotonergic system

A genetic contribution to the relationship between serotonin function and impulsivity/aggression is supported by many studies that demonstrate an influence exerted by various serotonin-related gene polymorphisms. Most studies refer to the serotonin transporter gene (Beitchman et al., 2003; Davidge et al., 2004; Hallikainen et al., 1999; Lee, Kim, & Hyun, 2003; Retz, Retz-Junginger, Supprian, Thome, & Rösler, 2004; Zalsman et al., 2001) and the tryptophan hydroxylase gene (Hennig, Reuter, Netter, Burk, & Landt, 2005; New et al., 1998; Rotondo et al., 1999; Rujescu, Giegling, Sato, Hartmann, & Möller, 2003; Staner et al., 2002; Zill et al., 2004). The enzyme tryptophan hydroxylase

(TPH) is involved in the synthesis of serotonin from the amino acid tryptophan, whereas the serotonin transporter (5-HTT) removes serotonin from the synaptic cleft. Another enzyme, monoamine oxidase-A (MAOA), selectively degrades serotonin, norepinephrine, and dopamine following reuptake from the synaptic cleft, and, therefore, plays a major role in the regulation of serotonin levels. A positive association between the allelic variant of the MAOA-uVNTR polymorphism coding for high MAOA activity and self-reported aggression and impulsivity in men was also reported (Manuck et al., 2000), which may be mediated, in part, by allel-specific variation in central serotonergic responsivity (Manuck et al., 2002). Sex-specific relationships between serotonin-related gene polymorphisms and impulsivity/aggression have been demonstrated for violent or clinical populations (Cadoret et al., 2003; Craig, Hart, Carson, McIlroy, & Passmore, 2004; New et al., 1998), as well as for healthy individuals (Manuck et al., 1999; Verona, Joiner, Johnson, & Bender, 2006). Again, effects were generally observed in men, but not in women. These findings may be indicative for a higher genotypic risk for impulsivity/aggression in males conferred by serotonin-related polymorphisms, as indicated in Figure 1 (middle column, link from serotonin-related polymorphisms to impulsivity, right column).

Caspi et al. (2002) were the first to demonstrate that a polymorphism in the MAOA gene moderates the capacity of experienced abuse in childhood to bring about the 'cycle of violence', where being maltreated as a child increases a person's risk for becoming the perpetrator of violence later in life. Childhood maltreatment led to increased adult violence only in those children, whose genotype conferred low levels of MAOA expression, while children with high activity MAOA genotype were protected against the negative effects of the same experience. This gene–environment interaction was confirmed by a replication study with a new sample of 975 7-year-old boys and a meta-analysis across four studies (Kim-Cohen et al., 2006). Although the moderating effect of the MAOA gene on violence was substantiated for males in these studies, it has been recently extended to both sexes, at least for white ethnicity samples (Widom & Brzustowicz, 2006). Gene–environment interactions are indicated in Figure 1 (reciprocal link between 'Environmental and Psychosocial Factors' and middle column with respect to serotonin-related gene polymorphisms).

4.3 Imaging genomics

Genes, however, even in the presence of toxic environments, do not directly code for violent behavior. Therefore, investigations of the neural mechanisms that might underlie the moderating genetic effects on (violent) behavior are needed. Hariri and Weinberger (2003) introduced 'imaging genomics' as an approach to study the influence of functional gene polymorphisms on brain activity. This technique allows for the estimation of genetic effects at the level of brain information processing, which represents a more proximate biological link between genes and behavior than correlations between genetic variation and behavioral phenotypes that are based on subjective experience. With regard to emotional behavior, imaging genomics has been used primarily to investigate the impact of serotonin-related gene polymorphisms on the functioning of the amygdala and prefrontal areas in response to fear, anxiety, and associated pathologies (Brown & Hariri, 2006; Hariri, Drabant, & Weinberger, 2006), whereas the investigation of impulsivity and aggression by means of imaging genomics is at the very beginning. The few existing studies focused on the MAOA genotype.

Passamonti et al. (2006) studied the influence of MAOA genetic variations on brain activity associated with impulse control in 24 healthy men by means of fMRI during a Go/NoGo task. The Barratt Impulsivity Scale (BIS) was used to assess trait impulsivity. The results showed a significantly greater response in the right lateral OFC for the high-activity allele carriers (MAOA-H) compared with the low-activity allele carriers (MAOA-L). BIS scores correlated positively with right lateral OFC activity in the MAOA-H carriers and negatively in the MAOA-L carriers. The authors concluded that genetic variation in the serotonergic system drive the brain response related to dispositional impulsivity. These findings raise the possibility that inconsistent results regarding the correlation between prefrontal activations and dispositional impulsivity as described in section 3.1 (Brown, Manuck, Flory, & Hariri, 2006; Horn, Dolan, Elliott, Deakin, & Woodruff, 2003) might be partially related to genetic variations in the MAOA gene. In support of this suggestion, the positive correlation between impulsivity and lateral OFC in MAOA-H carriers is consistent with the findings of Horn et al. (2003), whereas the negative association between impulsivity and lateral OFC in MAOA-L carriers fits the results of Brown et al. (2006) during the face matching task.

Meyer-Lindenberg et al. (2006) hypothesized that differences in the MAOA genotype may have an impact on structure and function of brain areas relevant for emotion regulation. They used structural MRI and different fMRI paradigms to activate the amygdala and regulatory frontal areas in healthy MAOA-L and MAOA-H individuals. The structural MRI results revealed that carriers of MAOA-L had volume reductions averaged around 8% relative to the volume in MAOA-H subjects in the rostral and dorsal ACC, amygdala, insula, and hypothalamus. The fMRI results showed increased amygdala activation and diminished reactivity of regulatory prefrontal areas (rostral ACC, OFC) for MAOA-L carriers compared with MAOA-H in a face-matching task. The authors postulate that their data identify differences in limbic circuitry for emotion regulation and cognitive control that may be involved in the association of MAOA with impulsive aggression. However, they also emphasize that their study was performed on psychiatrically normal volunteers, who were not characterized by increased levels of aggressive or violent behavior, so that they are not studying the relationship of MAOA and violence per se. Nonetheless, the work of Meyer-Lindenberg et al. (2006) is the first neuroimaging study that relates impulsive aggression to MAOA genotype-dependent reactivity of the neural circuitry of emotion regulation. Importantly, the findings of Meyer-Lindenberg et al. (2006) demonstrate a possible neural substrate of the gene–environment interaction observed by Caspi et al. (2002). It seems plausible to suggest that genetic risk factors act through brain regions implicated in the regulation of emotional behavior and modulated by serotonergic inputs to bring about individual differences in the vulnerability to environmental stress. However, it seems implausible to assume that the described differences in brain function are solely based on genetic variation in the MAOA gene. For example, amygdala hyperreactivity to emotionally provocative stimuli in normal volunteers has also been demonstrated as a consequence of genetic variation in the serotonin transporter (Hariri & Holmes, 2006), suggesting the need for considering gene-gene effects on emotion.

Eisenberger et al. (2007) examined in 32 healthy participants (19 female) how the MAOA polymorphism relates to self-reported aggression and interpersonal hypersensitivity, and to neural activity in response to an experimental episode of social exclusion. MAOA-L individuals were significantly higher in self-reported aggression and more interpersonally hypersensitive than MAOA-H individuals. Sex differences in these measures or the association of trait aggression and interpersonal hypersensitivity with the MAOA polymorphisms were not observed. The fMRI results during social exclusion showed greater activity of the dorsal ACC for the MAOA-L individuals. This finding is in contrast to the diminished dorsal ACC activity in MAOA-L carriers during perceptual matching of angry and fearful faces, as reported by Meyer-Lindenberg et al. (2006). This discrepancy might be explained by differences in stimulus characteristics. Probably, the social exclusion paradigm used by Eisenberger et al. (2007) elicited more intense negative feelings in MAOA-L individuals and, therefore, more regulatory engagement of the dorsal ACC than the perceptual matching task in the Meyer-Lindenberg et al. study.

Moreover, Eisenberger et al. (2007) conducted mediation analyses and found that the relationship between MAOA and trait aggression was mediated by both the dorsal ACC activity and trait interpersonal hyperactivity. The authors concluded that the association of MAOA genotype with aggression might be due to a heightened sensitivity to negative socioemotional experiences like social rejection, which characterizes MAOA-related aggression as impulsive (reactive) rather than instrumental. In this context, it might be interesting to note that social exclusion is a characteristic of indirect (also labeled 'relational' or 'social') aggression, which has been shown to be used more frequently by women than by men (Archer & Coyne, 2005; Björkqvist, Lagerspetz, & Kaukiainen, 1992; Crick & Grotpeter, 1995). From this it might be hypothesized that females are emotionally more affected by social exclusion than men, which should be reflected in a greater engagement of brain areas related to emotional pain in females compared to males. In fact, Eisenberger et al. (2007) mentioned that women showed more dorsal ACC activity than men, although this difference missed significance. In addition, no sex differences in the relationship between dorsal ACC activity and MAOA genotype were found. This lack of sex-related genotype effects in the study by Eisenberger et al. (2007) might be due to the small sample size (19 women, 13 men). Sample size is especially important when studying sex differences in relation to the MAOA-gene, because it is localized on the X-chromosome. Thus, males are hemizygous carriers of either one MAOA-L or MAOA-H allele, whereas females can be homozygous for each allele (MAOA-LL vs. MAOA-HH) or heterozygous (MAOA-LH). Eisenberger et al. (2007) reported that 10 of the 19 women were heterozygotes, thus, leaving approximately five females in each homozygous allele category for a direct comparison with their hemizygous male counterparts (the exact number of men and women in the MAOA-L and MAOA-H group is not given in the study). It is, therefore, possible that reduced statistical power prevented the detection of gene-by-sex interactions.

Meyer-Lindenberg et al. (2006) also investigated whether the variation of the MAOA genotype operates differently in men and women. With regard to brain structure, men, but not women, exhibited genotype-dependent changes in OFC volume, with MAOA-L men having approximately 14% more bilateral OFC volume than MAOA-H men. This finding is somewhat surprising, since higher OFC volume would be more consistent with better rather than insufficient emotion regulation capacities, contrary to what is associated with MAOA-L men. Possibly, this finding is influenced by the uneven male sample size with half the number of MAOA-L men ($n = 14$) compared to MAOA-H men ($n = 28$) participating in the structural measurements. Genotype effects on brain function showed for MAOA-L men increased reactivity of the hippocampus and amygdala during aversive recall (determining aversive scenes as new or old), and deficient activation of the dorsal ACC during a task related to inhibitory control. These effects were not observed in women. These genotype-dependent effects on brain function for men are shown in Figure 1 (middle column, link for MAOA-L men from serotonin-related gene polymorphisms to the ACC and amygdala; + indicates increased amygdala reactivity, – indicates deficient activation of the dorsal ACC).

4.4 Summary and conclusion

Converging lines of evidence from normal and abnormal populations demonstrate a relationship between serotonin dysregulation and increased impulsivity/aggression. Importantly, relationships are typically stronger for men or non-existent in women, suggesting differential serotonergic functioning and correlates of serotonin between men and women. Sex differences in the mean rate of brain serotonin synthesis and in the binding capacity of certain serotonin receptors have been reported for several brain regions including regulatory structures such as the ACC and OFC. The sex differences in the functioning of the serotonergic

system might be related to serotonin–hormone interactions, most notably the influence of female sex hormones such as estradiol and progesterone. The relationship between serotonin function and impulsivity/aggression is influenced by various serotonin-related gene polymorphisms, and there is some evidence for a higher genotypic risk for impulsivity/aggression in males conferred by serotonin-related polymorphisms. Longitudinal analyses revealed gene–environment interactions by demonstrating that childhood maltreatment led to increased adult violence only in those children, whose genotype conferred low levels of MAOA expression. As a possible neural substrate underlying this gene–environment interaction a prefrontal–amygdala system has been proposed. Importantly, genotype effects on structure and function of this neural system, which is implicated in the regulation of emotional behavior, are present in men but not women. Given its role for impulse control and the documented sex differences on different functional levels, the serotonin system is suggested to play a decisive role in mediating sex differences in aggression.

5. HORMONAL ASPECTS OF IMPULSIVITY AND AGGRESSION

5.1 Testosterone

The relationship between the steroid hormone testosterone and aggression has been of interest to researchers for a long time. A recent meta-analysis of 45 studies reported a weak positive correlation ($r = 0.14$) between testosterone and aggression (Book, Starzyk, & Quinsey, 2001). In a re-analysis of the same data with more stringent methodological criteria, an even lower but still significant overall association ($r = 0.08$) was found (Archer, Graham-Kevan, & Davies, 2005). This meta-analysis included 41 samples (range of r values: –0.41 to 0.75) of both sexes. Interestingly, the association was not restricted to men – in fact, it tended to be higher among women. This meta-analytic finding is in line with individual studies showing that higher levels of testosterone and other androgens in women are related to increased self-reported aggression (von der Pahlen, Lindman, Sarkola, Makisalo, & Eriksson, 2002), and an increased tendency to express competitive feelings through verbal aggression (Cashdan, 2003). Van Honk et al. (1999) found that testosterone levels were significantly related to negative affective states of anger and tension in both men and women. Bjork et al. (2001) reported that testosterone levels were highly correlated with a laboratory measure of impulsivity in a sample of young women. Coccaro et al. (2007b) assessed the relationship between testosterone, aggression, self-reported impulsivity, and venturesomeness in men with personality disorder. A significant positive correlation was found for venturesomeness only, which represents a form of sensation seeking. High testosterone levels have also been found in prison populations of both sexes, including violent criminals with antisocial personality disorder (Aromäki, Lindman, & Eriksson, 1999; Dabbs, Ruback, Frady, Hopper, & Sgoutas, 1988; Dabbs & Hargrove, 1997), and high testosterone has been related to the antisocial behavior factor in psychopathy (Stalenheim, Eriksson, von Knorring, & Wide, 1998).

A possible link between testosterone and some core symptoms of psychopathy is corroborated by studies examining short-term effects of a single dose of testosterone on aspects of motivation and emotion in healthy females. In this context, it was demonstrated that sublingual administration of testosterone reduces fear, as indicated by diminished fear-potentiated startle in a verbal threat-of-shock paradigm (Hermans, Putman, Baas, Koppeschaar, & van Honk, 2006), and empathy, as indexed by decreased facial expression mimicry in response to dynamic facial expressions of happy and angry faces (Hermans, Putman, & van Honk, 2006). A lack of empathy and fear is a characteristic determinant of a psychopathic personality (Hare, 2001). Moreover, administration of testosterone led to a shift in decision making to more risky choices in a gambling task, indicating a reduction in punishment sensitivity and heightened reward dependency (van Honk et al., 2004). Importantly, similar patterns of deficient decision making during this gambling task have been demonstrated in both clinical (Mitchell et al., 2002) and sub-clinical psychopaths (van Honk, Hermans, Putman, Montagne, & Schutter, 2002), and in patients with lesions of the OFC/VMPFC and presenting acquired sociopathy (Damasio, 1994). Taken together, the presented findings point to an association between testosterone and a number of aggression-related personality traits such as sensation seeking, impulsivity, negative emotionality, and empathy (see section 3.4). These relations are shown in Figure 1 (linking sex hormones, middle column, with personality traits, right column; + indicates positive association, – indicates negative association).

5.2 Cortisol

A positive relationship between testosterone and impulsive overt aggression was found in delinquent male adolescents (Popma et al., 2007). However, this relationship was specific to subjects with low cortisol levels and not apparent in subjects with high cortisol levels. In addition, no interaction effect was observed for a more controlled kind of aggression (covert aggression), but an inverse association between covert aggression and cortisol was found. These results indicate a moderating effect of cortisol on the association between testosterone and impulsive overt aggression, whereas more controlled covert aggression relates to low levels of cortisol.

O'Leary, Loney, and Eckel (2007) reported that psychopathic personality traits were related to cortisol responses to induced stress in a mixed-gender sample of college students. An association between high scores on self-reported psychopathic personality traits and lacking stress-induced cortisol increase was found for male, but not female participants. These findings suggest that cortisol production is a sex-specific marker for psychopathic personality traits. Loney et al. (2006) recruited a mixed-gender adolescent sample with various combinations of psychopathic personality traits and conduct problems. Whereas testosterone levels were found unrelated to psychopathic personality traits across male and female participants, resting cortisol production was inversely correlated with psychopathic personality traits. Again, this relationship was present in male, but not female adolescents. These findings indicate an association between a reduced response to stress and psychopathic personality traits in the normal range, which is in line with a hypofunction of the basic threat system found in criminal psychopaths (see section 2.3).

Because of the high moment-to-moment variation in adrenocortical activity, Shirtcliff, Granger, Booth, and Johnson (2005) proposed an analytical strategy in studying cortisol effects that separates variance in cortisol levels attributable to 'stable traitlike' from 'state or situationally specific' sources. Accordingly, cortisol levels were obtained from children and teenagers aged 6–16 on two successive days 1 year apart. For boys only, a negative correlation between the stable cortisol level and the degree of externalizing behavior was found. Collectively, these studies suggest that low cortisol is associated with heightened levels of covert aggression, externalizing behavior, and psychopathic traits (e.g., lack of fear and empathy) in males. This pattern of cortisol-related results parallel findings that demonstrate a relationship between frontal-limbic dysfunctions, antisocial behavior, and psychopathy, as reviewed in section 2. Since frontal-limbic structures are involved in the regulation of the hypothalamic–pituitary–adrenal (HPA) axis, which produces cortisol, results on the hormonal and on the brain level might be related. More specifically, the amygdala causes the hypothalamus to release corticotropin-releasing hormone (CRH), which activates the pituitary gland and promotes the release of adrenocorticotropic hormone (ACTH). ACTH, in turn stimulates cortisol release from the adrenal gland (De Bellis, 2005). In effect, the release of cortisol depends on the amygdala which is in turn modulated by prefrontal serotonergic input. This connection is indicated in Figure 1 (link from the amygdala to the HPA axis and cortisol, middle column).

The regulatory function of frontal-limbic structures holds also for the autonomic nervous system (ANS) (Zahn, Grafman, & Tranel, 1999), which relates to findings of an interaction between ANS and HPA activity in relation to aggressive behavior. Gordis et al. (2006) reported a correlation between increased aggression and low cortisol response in adolescents during stress-induction in a laboratory setting, but only when the ANS activity was low as well. Popma et al. (2006) compared HPA and ANS reactivity to stress in 12- and 14-year-old delinquent boys with and without disruptive behavior disorder (DBD). Compared to a normal control group, a decreased cortisol and heart rate response was found for delinquent boys with DBD, but not for those without DBD. This finding is in line with observations from children with DBD, exhibiting physiological underactivity including low cortisol levels (van Goozen & Fairchild, 2006). With respect to the ANS, Raine and colleagues (Raine, Venables, & Williams, 1990; Raine, Venables, & Mednick, 1997) described lower heart rates and lower skin conductance as predictors for later delinquency or aggression. A negative correlation between heart rate and aggression was confirmed for children aged 1–11 years in a meta-analysis by Lorber (2004). These findings suggest that hypofunction of the neurobiological stress systems (ANS, HPA axis) and, thus, a reduced stress reactivity increases the risk for habitual antisocial and aggressive behavior. However, a hyperresponsiveness of the stress systems also promotes aggressive behavior, which is consistent with findings of increased impulsive aggression in patients demonstrating OFC hypofunction and/or amygdala hyperresponsiveness (see section 4.2). Hyperarousal-related aggression has been related to an excessive acute glucocorticoid stress response, typically found in patients with intermittent explosive disorder, posttraumatic stress disorder, depression, and borderline personality disorder, whereas hypoarousal-related aggression is associated with glucocorticoid deficits that affect brain function on the long term, characteristic to conduct disorder and antisocial personality disorder (Haller, Mikics, Halasz, & Toth, 2005).

5.3 Summary and conclusion

Meta-analyses report a modest positive association between testosterone and aggression for both sexes. However, a number of recent studies provide evidence for a relationship between testosterone and aggression-related personality traits that have been shown to be sex-related in section 3.4. Cortisol seems to moderate the relationship between impulsive aggression and testosterone in delinquent male adolescents, since this relationship is found in individuals with low but not high cortisol levels. Low cortisol is also related to externalizing behavior and psychopathic personality traits in males but not females. Whereas low cortisol levels reflect HPA axis functioning, there are also indications for a reduced ANS activity in antisocial and aggressive populations. Together, these findings point to a reduced responsiveness to stress as risk factor for aggressive behavior that might be related to dysfunctions of frontal-limbic structures. This kind of hypoarousal-associated aggression has been related to antisocial personality disorder and its childhood precursor conduct disorder. A hypofunctional basic threat system is also characteristic of psychopathy (see sections 2.1 and 2.2) and might, therefore, relate to more instrumental aspects of aggression. Conversely, increased impulsive aggression relates to amygdala hyperresponsiveness and/or OFC hypofunction (see section 2.4), which is accompanied by an exaggerated response to stress, as seen for instance in intermittent explosive disorder and borderline personality disorder. Unfortunately, research on the relationship between stress reactivity and aggression generally does not differentiate between different forms of aggression (with Popma et al., 2007 as an exception). It is, therefore, not clear whether a hyporesponsiveness to stress relates to instrumental aggression.

6. THE INFLUENCE OF EARLY STRESS ON BRAIN DEVELOPMENT AND AGGRESSIVE BEHAVIOR AND ITS INTERRELATION WITH GENETIC FACTORS

6.1 Early stress and brain development

Early stressful life events are multifold and include prenatal toxic influences such as maternal substance abuse during pregnancy, deprivation of normal parental care during infancy, childhood physical maltreatment, childhood neglect, and exposure to family conflict and violence. Exposure to early life maltreatment, in the form of childhood abuse and neglect, is a risk factor for emotional instability associated with several psychiatric disorders, and there is increasing evidence that adverse experience affects important neurodevelopmental processes, including neurogenesis, synaptic pruning, and myelination during specific, sensitive periods (De Bellis, 2005; Teicher et al., 2003).

During normal brain development, the final trimester of pregnancy and the postnatal period up to the fourth, fifth or sixth year are considered to be the phases of human life when the greatest brain growth occurs (Monk, Webb, & Nelson, 2001; Webb, Monk, & Nelson, 2001). The greatest degree of networking between the nerve cells of the brain takes place during this time; it is around three times higher than in adults. In a longitudinal MRI study, Giedd et al. (1999) reported a linear increase of the volume of white matter across ages 4–22, while cortical gray matter development was nonlinear, with a pre-adolescent increase and a decline during post-adolescence. Similarly, longitudinal imaging data by Gogtay et al. (2004) on gray matter development between the ages of 4 and 21 years revealed an increase of total gray matter volume at earlier ages, followed by sustained loss starting around puberty. This process of gray matter loss begins in the primary sensory areas, and then spreads rostrally over the frontal cortex, with the prefrontal and orbitofrontal cortex maturing last (Gogtay et al., 2004). These long-lasting maturational changes in the structure of brain areas associated with impulse inhibition and social behavior might also provide the basis for the impact of stress and other negative environmental influences on brain development.

Teicher et al. (2003) examined the influence of severe stress in childhood on brain development, and identified several brain regions including the hippocampus, the amygdala, and the prefrontal cortex as especially vulnerable to stress. The high stress vulnerability of these structures might be caused either by long-lasting maturation, typical of the pre- and orbitofrontal cortex, or by a high density of glucocorticoid (cortisol) receptors as is the case with the hippocampus, which is involved in stress regulation. The stress sensitivity of the amygdala is grounded in its tendency to increase its excitability with repeated intermittent stimulation, leading to 'limbic irritability' in the long term (Teicher et al., 2003). Given the strong interconnections between the OFC and amygdala, it was hypothesized that chronic amygdala activation in neglected children impairs the development of the prefrontal cortex leading to problems with the normal age-related acquisition of behavioral and emotional regulation including the inhibition of impulsive behaviors (De Bellis, 2005). This conceptualization of the effect of early stress on brain functioning is supported by neuroimaging findings in clinical populations characterized by increased levels of impulsivity/aggression and OFC hypofunction and/or amygdala hyperresponsiveness, as reviewed in section 2.4. The diagnoses of these patients included intermittent explosive disorder, borderline personality disorder, and suicidal behaviour (self-directed aggression) which have been shown to be related to serotonin dysregulation in prefrontal regions (see section 4.1). There are indications that normal functioning of the serotonin system is compromised by early life stress, and it has been proposed that early trauma interacts with diminished serotonin function, which may differentially lead to suicidal and aggressive behavior (Gollan et al., 2005). Furthermore, child maltreatment has complex, long-term influences on the functioning of the HPA axis (Tarullo & Gunnar, 2006). The HPA axis, the ANS, and the serotonin system are interconnected at many levels, such that maltreatment-related dysregulation in one system can lead to problems in others (Watts-English, Fortson, Gibler, Hooper, & De Bellis, 2006).

6.2 Gene–environment interaction

Although the evidence reviewed in section 6.1 highlight the enduring effects of adverse environmental conditions on the neurobiological stress systems, there are also child-specific factors that might predispose to antisocial and aggressive behavior. Longitudinal studies demonstrate developmental links from early childhood temperament to later adult personality traits that relate to aggression. For instance, undercontrolled children rated as irritable, impulsive, and emotionally labile at age 3, were intolerant, overreacting to minor events, and high scoring on traits indexing negative emotionality at age 26 (Caspi et al., 2003). This is exactly the personality profile that has been found to relate to a developmental trajectory of life-course persistent antisocial behavior (Moffitt, Caspi, Dickson, Silva, & Stanton, 1996), and that explained most of the sex-related variance in antisocial behavior (see section 3.4). This indicates that a personality style of low impulse control and high negative emotionality stays longitudinally consistent across adulthood and, thus, represents a predisposition for aggressive and antisocial behavior that may be expressed as antisocial personality disorder and its childhood precursor conduct disorder. However, whether such a maladaptive extreme emerges or not strongly depends on the presence of environmental risk factors and, more importantly, on the vulnerability of an individual to these risk factors, which differs between individuals according to their genetic outfit and might be reflected in different brain functioning (Caspi & Moffitt, 2006). It is, therefore, likely that the influence of early stress on brain development, stress system functioning, and the serotonergic system depends, in part, on individual genetic variation. An example of such a gene–environment interaction has been described in section 4.2, demonstrating that the influence of childhood maltreatment on adult violence is moderated by the MAOA genotype (Caspi et al., 2002; Kim-Cohen et al., 2006).

The concept of vulnerability has also been stressed in relation to antisocial and aggressive children with DBD, which are known to have been often exposed to neglect and abuse, but present with low stress reactivity (van Goozen & Fairchild, 2006), whereas maltreated children with internalizing disorders often show an increased responsiveness to stress (Tarullo & Gunnar, 2006). These differences are likely related to differences in child-specific factors that shape the temperament of a child and, in turn, its reaction to environmental stress. There is increasing evidence that variation in genetic polymorphisms contribute to individual differences in personality and temperament, including stress reactivity, through their influence on the functioning of the serotonin and other neurotransmitter systems (Munafò et al., 2003; Reif & Lesch, 2003). Furthermore, the temperament of a child influences the quality of its interaction with the social environment. For example, a child with a difficult temperament or problem behavior might elicit frequent negative responses from parents, siblings, and peers that in turn increase the risk for the child of being neglected. Depending on its biological outfit, the frequent exposure to stressful situations might result in a habituation to stress as indicated by low stress reactivity and an increased risk of developing externalizing behavior problems (van Goozen & Fairchild, 2006). Alternatively, the child might become sensitized to stress, leaving it with a greater risk of developing internalizing behavior problems, such as depression and anxiety disorders (Tarullo & Gunnar, 2006). This implies, on the other hand, that an advantageous social environment with a responsive care provider might serve to buffer the HPA system. In this context, it could be shown that children whose mothers react sensitively and appropriately showed a reduced stress reaction, whereas children who experienced low maternal responsiveness exhibited a higher cortisol response in a stressful situation (Gunnar, Brodersen, Nachmias, Buss, & Rigatuso, 1996). Taken together, the reported findings show that a complex interaction between environment, genes, personality, and brain function determines an individual's stress sensitivity and, related to this, the risk of developing externalizing and/or internalizing behavior problems. This interplay is depicted in Figure 1 by linking 'Environmental and Psychosocial factors' at the top with the three columns 'Neuronal Structures', 'Neurophysiology', and 'Personality'.

6.3 Summary and conclusions

Exposure to early life maltreatment is a risk factor for emotional instability associated with a number of psychiatric disorders. Childhood abuse and neglect affect the development of prefrontal-limbic structures, the serotonin system, and the HPA axis. The functioning of these systems is highly interrelated and significantly influences arousal, stress reactions, and emotional regulation. The effects induced by adverse environmental factors interact with child-specific temperamental factors that are, in part, under genetic control and reflect an individual's vulnerability to environmental risk factors. This gene–environment interaction defines how the brain reacts to stress later in life and whether there is an increased risk of developing externalizing and/or internalizing behavior problems. It is concluded that the interaction between early life stress and individual genetic factors is of great importance for a better understanding of the relationship between cortisol and aggression (see section 5.2).

7. GENERAL CONCLUSIONS

The purpose of this article was to identify neurobiological factors that may provide a better understanding of the mechanisms that underlie the association between antisocial and criminal behavior and low self-control/high impulsivity on the one hand, and the predominance of men in antisocial and criminal behaviors on the other. More specifically, we hypothesized that the concept of impulse control might serve to integrate criminological, psychological, clinical, and neurobiological explanations of aggressive behavior. In support of this hypothesis, we demonstrated that a relation between low impulse control and aggression exists on the clinical, neurobiological, and psychological level. Furthermore, we have outlined possible links between the different research fields by stepwise developing an integrative model of impulsive aggression that illustrates a complex interaction between environment, genes, neurophysiology, brain function, and personality (Figure 1).

According to this model, a personality style of high impulsivity and negative emotionality together with low empathy is neurobiologically characterized by relatively high levels of testosterone, HPA-axis function irregularities, a reduced functioning of the serotonergic system, and a diminished capacity of prefrontal brain structures in regulating amygdala activation. The neurobiological systems are interconnected at many levels, such that dysregulation in one system can lead to problems in others. During brain development, a complex gene–environment interaction shapes the interrelated functioning of these systems and, in turn, the personality style of an individual.

We had a special focus on sex differences throughout this review. Sex differences in aggression and violence are well established in the psychosocial and criminological literature, but neurobiological research on sex differences in human aggression is sparse. However, there are indications for sex differences in the functioning of the serotonergic system. Most studies that investigated a relationship between serotonin or serotonin-related polymorphisms and impulsivity/aggression found significant relations for men only. In addition, there is neuroanatomical evidence for a higher OFC-to-amygdala volume and an increased functional connectivity between OFC and amygdala in women. Together, neurobiological findings point to an increased capacity for women to control impulses and regulate emotions. As regards personality traits, we reviewed evidence that men score higher on personality dimensions that are positively correlated with aggression and negatively with behavioral inhibition, whereas women score higher on dimensions that are negatively associated with aggression and positively with behavioral inhibition. These sex differences in personality and brain function as well as genetic and neurophysiological factors are indexed in the model (Figure 1). Accordingly, we propose that sex-related differences in the neurobiological correlates of impulse control and emotion regulation underlie the sex differences in personality that in turn mediate the sex differences in aggression and antisocial behavior. Psychosocial factors such as gender roles and social norms, which restrict women more than men in 'acting out' aggressive feelings, exert their influence in the same direction as the neurobiological and personality factors. Therefore, these environmental factors might lead to a further increase of behavioral inhibition in women that mediates the sex difference in direct aggression.

Although the sex differences indicated in the model rely on findings in healthy normal individuals, the sex-related differences in various neurobiological aspects that are important for the regulation of emotion might have implications for the explanation of sex differences in the prevalence of psychiatric disorders. Most notably in the context of the present review, there is a male preponderance for externalizing behavior problems, such as aggressive and antisocial behavior (Moffitt, Caspi, Rutter, & Silva, 2001), and a higher prevalence of internalizing disorders such as depression in females (Cyranowski, Frank, Young, & Shear, 2000). We suggest that this sex difference also relates to the described interaction between genes, environment, brain function, and personality. Under adverse environmental conditions the normal sex-related variation in emotion regulation mechanisms leave males with a higher risk of 'acting out' aggressively, whereas females show a higher risk for anxiety-related disorders or self-directed aggression, as is the case with borderline personality disorder.

As a caveat, this article was not meant to present a critical and comprehensive review of all available literature or all possible variables that may contribute to aggressive behavior and sex differences in aggression. Rather, we intended to outline a hypothetical framework that identifies impulse control as an integrative concept for studying sex differences in aggression and antisocial behavior. Therefore, the presented model might serve as a heuristic tool and stimulate integrative research on this understudied topic.

REFERENCES

Adams, R. B. Jr., Gordon, H. L., Baird, A. A., Ambady, N., & Kleck, R. E. (2003). Effects of gaze on amygdala sensitivity to anger and fear faces. *Science, 300,* 1536.

Anderson, S. W., Bechara, A., Damasio, H., Tranel, D., & Damasio, A. R. (1999). Impairment of social and moral behavior related to early damage in human prefrontal cortex. *Nature Neuroscience, 2,* 1032–1037.

Arango, V., Underwood, M. D., Gubbi, A. V., & Mann, J. J. (1995). Localized alterations in pre- and postsynaptic serotonin binding sites in the ventrolateral prefrontal cortex of suicide victims. *Brain Research, 688,* 121–133.

Archer, J. (2004). Sex differences in aggression in real-world settings: A meta-analytic review. *Review of General Psychology, 8,* 291–322.

Archer, J., & Coyne, S. M. (2005). An integrated review of indirect, relational, and social aggression. *Personality and Social Psychology Review, 9,* 212–230.

Archer, J., Graham-Kevan, N., & Davies, M. (2005). Testosteron and aggression: A re-analysis of Book, Starzyk and Quinsey's (2001) study. *Aggression and Violent Behavior, 10,* 241–261.

Aromäki, A. S., Lindman, R. E., & Eriksson, C. J. P. (1999). Testosterone, aggressiveness, and antisocial personality. *Aggressive Behavior, 25,* 113–123.

American Psychiatric Association, A. P. (1994). *Diagnostic and Statistical Manual of Mental Disorders* (4th ed.). Washington, DC: American Psychiatric Association.

Baron-Cohen, S., Knickmeyer, R. C., & Belmonte, M. K. (2005). Sex differences in the brain: Implications for explaining autism. *Science, 310,* 819–823.

Barratt, E. S., Stanford, M. S., Dowdy, L., Liebman, M. J., & Kent, T. A. (1999). Impulsive and premeditated aggression: A factor analysis of self-reported acts. *Psychiatry Research, 86,* 163–173.

Bassarath, L. (2001). Neuroimaging studies of antisocial behaviour. *Canadian Journal of Psychiatry, 46,* 728–732.

Bechara, A. (2004). The role of emotion in decision-making: Evidence from neurological patients with orbitofrontal damage. *Brain and Cognition, 55,* 30–40.

Bechara, A., Damasio, H., Tranel, D., & Damasio, A. R. (1997). Deciding advantageously before knowing the advantageous strategy. *Science, 275,* 1293–1295.

Beitchman, J. H., Davidge, K. M., Kennedy, J. L., Atkinson, L., Lee, V., Shapiro, S., & Douglas, L. (2003). The serotonin transproter gene in aggressive children with and without ADHD and nonaggressive matched controls. *Annals of the New York Academy of Sciences, 1008,* 248–251.

Berlin, H. A., Rolls, E. T., & Kischka, U. (2004). Impulsivity, time perception, emotion and reinforcement sensitivity in patients with orbitofrontal cortex lesions. *Brain, 127,* 1108–1126.

Best, M., Williams, J. M., & Coccaro, E. F. (2002). Evidence for a dysfunctional prefrontal circuit in patients with an impulsive aggressive disorder. *Proceedings of the National Academy of Sciences of the United States of America, 99,* 8448–8453.

Bettencourt, B. A., & Miller, N. (1996). Gender differences in aggression as a function of provocation: A meta-analysis. *Psychological Bulletin, 119,* 422–447.

Birbaumer, N., Veit, R., Lotze, M., Erb, M., Hermann, C., Grodd, W., & Flor, H. (2005). Deficient fear conditioning in psychopathy: A functional magnetic resonance imaging study. *Archives of General Psychiatry, 62,* 799–805.

Biver, F., Lotstra, F., Monclus, M., Damhaut, P., Mendlewicz, J., & Goldman, S. (1996). Sex difference in 5HT2 receptor in the living human brain. *Neuroscience Letters, 204,* 25–28.

Bjork, J. M., Dougherty, D. M., Moeller, F. G., & Swann, A. C. (2000). Differential behavioral effects of plasma tryptophan depletion and loading in aggressive and nonaggressive men. *Neuropsychopharmacology, 22,* 357–369.

Bjork, J. M., Moeller, F. G., Dougherty, D. M., & Swann, A. C. (2001). Endogenous plasma testosterone levels and commission errors in women: A preliminary report. *Physiology & Behavior, 73,* 217–221.

Björkqvist, K., Lagerspetz, K. M. J., & Kaukiainen, A. (1992). Do girls manipulate and boys fight? Developmental trends regarding direct and indirect aggression. *Aggressive Behavior, 18,* 117–127.

Blair, K. S., Newman, C., Mitchell, D. G. V., Richell, R. A., Leonard, A., Morton, J., & Blair, R. J. R. (2006). Differentiating among prefrontal substrates in psychopathy: Neuropsychological test findings. *Neuropsychology, 20,* 153–165.

Blair, R. J. R. (2001). Neurocognitive models of aggression, the antisocial personality disorders, and psychopathy. *Journal of Neurology, Neurosurgery & Psychiatry, 71,* 727–731.

Blair, R. J. R. (2003). Neurobiological basis of psychopathy. *British Journal of Psychiatry, 182,* 5–7.

Blair, R. J. R. (2004). The roles of orbital frontal cortex in the modulation of antisocial behavior. *Brain and Cognition, 55,* 198–208.

Blair, R. J. R. (2005). Responding to the emotions of others: Dissociating forms of empathy through the study of typical and psychiatric populations. *Consciousness and Cognition, 14,* 698–718.

Blair, R. J. R. (2006). The emergence of psychopathy: Implications for the neuropsychological approach to developmental disorders. *Cognition, 101,* 414–442.

Blair, R. J. R., & Cipolotti, L. (2000). Impaired social response reversal: A case of 'acquired sociopathy'. *Brain, 123,* 1122–1141.

Blair, R. J. R., Morris, J. S., Frith, C. D., Perrett, D. I., & Dolan, R. J. (1999). Dissociable neural responses to facial expressions of sadness and anger. *Brain, 122,* 883–893.

Blair, R. J. R., Colledge, E., & Mitchell, D. G. V. (2001). Somatic markers and response reversal: Is there orbitofrontal cortex dysfunction in boys with psychopathic tendencies? *Journal of Abnormal Child Psychology, 29,* 499–511.

Blair, R. J. R., Colledge, E., Murray, L., & Mitchell, D. G. V. (2001). A selective impairment in the processing of sad and fearful expressions in children with psychopathic tendencies. *Journal of Abnormal Child Psychology, 29,* 491–498.

Blair, R. J. R., Peschardt, K. S., Budhani, S., Mitchell, D. G. V., & Pine, D. S. (2006). The development of psychopathy. *Journal of Child Psychology and Psychiatry, 47,* 262–276.

Blonigen, D. M., Carlson, S. R., Krueger, R. F., & Patrick, C. J. (2003). A twin study of self-reported psychopathic personality traits. *Personality and Individual Differences, 35,* 179–197.

Bond, A. J., Wingrove, J., & Critchlow, D. G. (2001). Tryptophan depletion increases aggression in women during the premenstrual phase. *Psychopharmacology, 156,* 477–480.

Bonne, O., Grillon, C., Vythilingam, M., Neumeister, A., & Charney, D. S. (2004). Adaptive and maladaptive psychobiological responses to severe psychological stress: Implications for the discovery of novel pharmacotherapy. *Neuroscience and Biobehavioral Reviews, 28,* 65–94.

Book, A. S., Starzyk, K. B., & Quinsey, V. L. (2001). The relationship between testosterone and aggression: A meta-analysis. *Aggression and Violent Behavior, 6,* 579–599.

Brower, M. C., & Price, B. H. (2001). Neuropsychiatry of frontal lobe dysfunction in violent and criminal behaviour: A critical review. *Journal of Neurology, Neurosurgery & Psychiatry, 71,* 720–726.

Brown, S. M., & Hariri, A. R. (2006). Neuroimaging studies of serotonin gene polymorphisms: Exploring the interplay of genes, brain, and behavior. *Cognitive, Affective, & Behavioral Neuroscience, 6,* 44–52.

Brown, S. M., Manuck, S. B., Flory, J. D., & Hariri, A. R. (2006). Neural basis of individual differences in impulsivity: Contributions of corticolimbic circuits for behavioral arousal and control. *Emotion, 6,* 239–245.

Budhani, S., Richell, R. A., & Blair, R. J. R. (2006). Impaired reversal but intact acquisition: Probabilistic response reversal deficits in adult individuals with psychopathy. *Journal of Abnormal Psychology, 115,* 552–558.

Bufkin, J. L., & Luttrell, V. R. (2005). Neuroimaging studies of aggressive and violent behavior. Current findings and implications for criminology and criminal justice. *Trauma, Violence, & Abuse, 6,* 176–191.

Burton, V. S., Cullen, F. T., Evans, T. D., Alardi, L. F., & Dunaway, R. G. (1998). Gender, self-control and crime. *Journal of Research in Crime and Delinquency, 35,* 123–147.

Cadoret, R. J., Langbehn, D., Caspers, K., Troughton, E. P., Yucuis, R., Sandhu, H. K., & Philibert, R. (2003). Associations of the serotonin transporter promoter polymorphism with aggressivity, attention deficit, and conduct disorder in an adoptee population. *Comprehensive Psychiatry, 44,* 88–101.

Campbell, A. (2006). Sex differences in direct aggression: What are the psychological mediators? *Aggression and Violent Behavior, 11,* 237–264.

Campbell, A., & Muncer, S. (1994). Sex differences in aggression: Social roles and social representations. *British Journal of Social Psychology, 33,* 233–240.

Cashdan, E. (2003). Hormones and competitive aggression in women. *Aggressive Behavior, 29,* 107–115.

Caspi, A., & Moffitt, T. E. (2006). Gene-environment interactions in psychiatry: Joining forces with neuroscience. *Nature Reviews Neuroscience, 7,* 583–590.

Caspi, A., McClay, J., Moffitt, T. E., Mill, J., Martin, J., Craig, I. W., Taylor, A., & Poulton, R. (2002). Role of genotype in the cycle of violence in maltreated children. *Science, 297,* 851–854.

Caspi, A., Harrington, H., Milne, B., Amell, J. W., Theodore, R. F., & Moffitt, T. E. (2003). Children's behavioral styles at age 3 are linked to their adult personality traits at age 26. *Journal of Personality, 71,* 495–513.

Cleare, A. J., & Bond, A. J. (1997). Does central serotonergic function correlate inversely with aggression? A study using d-fenfluramine in healthy subjects. *Psychiatry Research, 69,* 89–95.

Cleare, A. J., & Bond, A. J. (2000). Ipsapirone challenge in aggressive men shows an inverse correlation between 5-HT$_{1A}$ receptor function and aggression. *Psychopharmacology, 148,* 344–349.

Coccaro, E. F. (2003). Intermittent explosive disorder. *Current Psychiatry, 2,* 42–60.

Coccaro, E. F., Gabriel, S., & Siever, L. J. (1990). Buspirone challenge: Preliminary evidence for a role for central 5-HT1a receptor function in impulsive aggressive behavior in humans. *Psychopharmacology Bulletin, 26,* 393–405.

Coccaro, E. F., Beresford, B., Minar, P., Kaskow, J., & Geracioti, T. (2007a). CSF testosterone: Relationship to aggression, impulsivity, and venturesomeness in adult males with personality disorder. *Journal of Psychiatric Research, 41,* 488–492.

Coccaro, E. F., McCloskey, M. S., Fitzgerald, D. A., & Phan, K. L. (2007b). Amygdala and orbitofrontal reactivity to social threat in individuals with impulsive aggression. *Biological Psychiatry, 62,* 168–178.

Craig, D., Hart, D. J., Carson, R., McIlroy, S. P., & Passmore, A. P. (2004). Allelic variation at the A218C tryptophan hydroxylase polymorphism influences agitation and aggression in Alzheimer's disease. *Neuroscience Letters, 363,* 199–202.

Crick, N. R., & Grotpeter, J. K. (1995). Relational aggression, gender, and social-psychological adjustment. *Child Development, 66,* 710–722.

Cyranowski, J. M., Frank, E., Young, E., & Shear, M. K. (2000). Adolescent onset of the gender difference in lifetime rates of major depression. *Archives of General Psychiatry, 57,* 21–27.

Dabbs Jr., J. M., & Hargrove, M. F. (1997). Age, testosterone, and behavior among female prison inmates. *Psychosomatic Medicine, 59,* 477–480.

Dabbs, J. M., Ruback, R. B., Frady, R. L., Hopper, D. H., & Sgoutas, D. S. (1988). Saliva testosterone and criminal violence among women. *Personality and Individual Differences, 9,* 269–275.

Damasio, A. R. (1994). *Descartes' error: Emotion, rationality and the human brain.* New York: Putnam (Grosset Books).

Damasio, A. R., Grabowski, T. J., Bechara, A., Damasio, H., Ponto, L. L. B., Parvizi, J., & Hichwa, V. (2000). Subcortical and cortical brain activity during the feeling of self-generated emotions. *Nature Neuroscience, 3,* 1049–1055.

Davidge, K. M., Atkinson, L., Douglas, L., Lee, V., Shapiro, S., Kennedy, J. L., & Beitchman, J. H. (2004). Association of the serotonin transporter and 5HT1D beta receptor genes with extreme, persistent and pervasive aggressive behaviour in children. *Psychiatric Genetics, 14,* 143–146.

Davidson, R. J., Putnam, K. M., & Larson, C. L. (2000). Dysfunction in the neural circuitry of emotion regulation– A possible prelude to violence. *Science, 289,* 591–594.

De Bellis, M. D. (2005). The psychobiology of neglect. *Child Maltreatment, 10,* 150–172.

Deeley, Q., Daly, E., Surguladze, S., Tunstall, N., Mezey, G., Beer, D., Ambikapathy, A., Robertson, D., Giampietro, V., Brammer, M. J., Clarke, A., Dowsett, J., Fahy, T. O. M., Phillips, M. L., & Murphy, D. G. (2006). Facial emotion processing in criminal psychopathy: Preliminary functional magnetic resonance imaging study. *British Journal of Psychiatry, 189,* 533–539.

Dougherty, D. D., Shin, L. M., Alpert, N. M., Pitman, R. K., Orr, S. P., Lasko, M., Macklin, M. L., Fischman, A. J., & Rauch, S. L. (1999a). Anger in healthy men: A PET study using script-driven imagery. *Biological Psychiatry, 46,* 466–472.

Dougherty, D. M., Bjork, J. M., Marsh, D. M., & Moeller, F. G. (1999b). Influence of trait hostility on tryptophan depletion-induced laboratory aggression. *Psychiatry Research, 88,* 227–232.

Dougherty, D. D., Rauch, S. L., Deckersbach, T., Marci, C., Loh, R., Shin, L. M., Alpert, N. M., Fischman, A. J., & Fava, M. (2004). Ventromedial prefrontal cortex and amygdala dysfunction during an anger induction positron emission tomography study in patients with major depressive disorder with anger attacks. *Archives of General Psychiatry, 61,* 795–804.

Driscoll, H., Zinkivskay, A., Evans, K., & Campbell, A. (2006). Gender differences in social representations of aggression: The phenomenological experience of differences in inhibitory control? *British Journal of Psychology, 97,* 139–153.

Eagly, A. H., & Steffen, V. J. (1986). Gender and aggressive behavior: A meta-analytic review of the social psychological literature. *Psychological Bulletin, 100,* 309–330.

Eisenberger, N. I., Way, B. M., Taylor, S. E., Welch, W. T., & Lieberman, M. D. (2007). Understanding genetic risk for aggression: Clues from the brain's response to social exclusion. *Biological Psychiatry, 61,* 1100–1108.

Fitzgerald, D. A., Angstadt, M., Jelsone, L. M., Nathan, P. J., & Phan, K. L. (2006). Beyond threat: Amygdala reactivity across multiple expressions of facial affect. *NeuroImage, 30,* 1441–1448.

Flory, J. D., Harvey, P. D., Mitropoulou, V., New, A. S., Silverman, J. M., Siever, L. J., & Manuck, S. B. (2006). Dispositional impulsivity in normal and abnormal samples. *Journal of Psychiatric Research, 40,* 438–447.

Frankle, W. G., Lombardo, I., New, A. S., Goodman, M., Talbot, P. S., Huang, Y., Hwang, D.-R., Slifstein, M., Curry, S., Abi-Dargham, A., Laruelle, M., & Siever, L. J. (2005). Brain serotonin transporter distribution in subjects with impulsive aggressivity: A positron emission study with [11C]McN 5652. *American Journal of Psychiatry, 162,* 915–923.

Giedd, J. N., Blumenthal, J., Jeffries, N. O., Castellanos, F. X., Liu, H., Zijdenbos, A., Paus, T., Evans, A. C., & Rapoport, J. L. (1999). Brain development during childhood and adolescence: A longitudinal MRI study. *Nature Neuroscience, 2,* 861–863.

Gogtay, N., Giedd, J. N., Luisk, L., Hayashi, K. M., Greenstein, D., Vaituzis, A. C., Nugent, T. F., Herman, D. H., Clasen, L. S., Taga, A. W., Rapoport, J. L., & Thompson, P. M. (2004). Dynamic mapping of human cortical development during childhood through early adulthood. *Proceedings of the National Academy of Sciences of the United States of America, 101,* 8174–8179.

Gollan, J. K., Lee, R., & Coccaro, E. F. (2005). Developmental psychopathology and neurobiology of aggression. *Development and Psychopathology, 17,* 1151–1171.

Gordis, E. B., Granger, D. A., Susman, E. J., & Trickett, P. K. (2006). Asymmetry between salivary cortisol and alpha-amylase reactivity to stress: Relation to aggressive behavior in adolescents. *Psychoneuroendocrinology, 31,* 976–987.

Gottfredson, M. R., & Hirschi, T. (1990). *A general theory of crime.* Stanford: University Press.

Grafman, J., Schwab, K., Warden, D. R., Pridgen, A., Brown, H. R., & Salazar, A. M. (1996). Frontal lobe injuries, violence, and aggression: A report of the Vietnam Head Injury Study. *Neurology, 46,* 1231–1238.

Gunnar, M. R., Brodersen, L., Nachmias, M., Buss, K., & Rigatuso, J. (1996). Stress reactivity and attachment security. *Developmental Psychobiology, 29,* 191–204.

Gur, R. C., Gunning-Dixon, F., Bilker, W. B., & Gur, R. E. (2002). Sex differences in temporo-limbic and frontal brain volumes of healthy adults. *Cerebral Cortex, 12,* 998–1003.

Haller, J., Mikics, E., Halasz, J., & Toth, M. (2005). Mechanisms differentiating normal from abnormal aggression: Glucocorticoids and serotonin. *European Journal of Pharmacology, 526,* 89–100.

Hallikainen, T., Saito, T., Lachman, H. M., Volavka, J., Pohjalainen, T., Ryynänen, O.-P., Kauhanen, J., Syvälahti, E., Hietala, J., & Tiihonen, J. (1999). Association between low activity serotonin transporter promoter genotype and early onset alcoholism with habitual impulsive violent behavior. *Molecular Psychiatry, 4,* 385–388.

Hamann, S., & Canli, T. (2004). Individual differences in emotion processing. *Current Opinion in Neurobiology, 14,* 1–6.

Hare, R. D. (1991). *The Hare Psychopathy Checklist-Revised.* Toronto, Ontario: Multi-Health Systems.

Hare, R. D. (2001). Psychopaths and their nature. In A. Raine & J. Sanmartin (Eds), *Violence and psychopathy* (pp. 5–34). New York: Kluwer Academic/Plenum Publishers.

Hare, R. D., Clark, D., Grann, M., & Thornton, D. (2000). Psychopathy and the predictive validity of the PCL-R: An international perspective. *Behavioral Sciences & the Law, 18,* 623–645.

Hariri, A., & Holmes, A. (2006). Genetics of emotional regulation: The role of the serotonin transporter in neural function. *Trends in Cognitive Sciences, 10,* 182–191.

Hariri, A. R., & Weinberger, D. R. (2003). Imaging genomics. *British Medical Bulletin, 65,* 259–270.

Hariri, A., Drabant, E. M., & Weinberger, D. R. (2006). Imaging genetics: Perspectives from studies of genetically driven variation in serotonin function and corticolimbic affective processing. *Biological Psychiatry, 59,* 888–897.

Harpur, T. J., Hare, R. D., & Hakstian, A. R. (1989). Two-factor conceptualization of psychopathy: Construct validity and assessment implications. *Psychological Assessment, 1,* 6–17.

Hennig, J., Reuter, M., Netter, P., Burk, C., & Landt, O. (2005). Two types of aggression are differentially related to serotonergic activity and the A779C TPH polymorphism. *Behavioral Neuroscience, 119,* 16–25.

Hermans, E. J., Putman, P., Baas, J. M., Koppeschaar, H. P., & van Honk, J. (2006). A single administration of testosterone reduces fear-potentiated startle in humans. *Biological Psychiatry, 59,* 872–874.

Hermans, E. J., Putman, P., & van Honk, J. (2006). Testosterone administration reduces empathetic behavior: A facial mimicry study. *Psychoneuroendocrinology, 31,* 859–866.

Herpertz, S. C., Werth, U., Lukas, G., Qunaibi, M., Schuerkens, A., Kunert, H. J., Freese, R., Flesch, M., Mueller-Isberner, R., Osterheider, M., & Sass, H. (2001). Emotion in criminal offenders with psychopathy and borderline personality disorder. *Archives of General Psychiatry, 58,* 737–745.

Hoptman, M. J., Volavka, J., Johnson, G., Weiss, E., Bilder, R. M., & Lim, K. O. (2002). Frontal white matter microstructure, aggression, and impulsivity in men with schizophrenia: A preliminary study. *Biological Psychiatry, 52,* 9–14.

Horn, N. R., Dolan, M., Elliott, R., Deakin, J. F. W., & Woodruff, P. W. R. (2003). Response inhibition and impulsivity: An fMRI study. *Neuropsychologia, 41,* 1959–1966.

Hornak, J., Bramham, J., Rolls, E. T., Morris, R. G., O'Doherty, J., Bullock, P. R., & Polkey, C. E. (2003). Changes in emotion after circumscribed surgical lesions of the orbitofrontal and cingulate cortices. *Brain, 126,* 1691–1712.

Ishikawa, S. S., Raine, A., Lencz, T., Bihrle, S., & LaCasse, L. (2001). Autonomic stress reactivity and executive functions in successful and unsuccessful criminal psychopaths from the community. *Journal of Abnormal Psychology, 110,* 423–432.

Kampe, K. K. W., Frith, C. D., & Frith, U. (2003). 'Hey John': Signals conveying communicative intention toward the self activate brain regions associated with 'mentalizing,' regardless of modality. *Journal of Neuroscience, 23,* 5258–5263.

Kiehl, K. A. (2006). A cognitive neuroscience perspective on psychopathy: Evidence for paralimbic system dysfunction. *Psychiatry Research, 142,* 107–128.

Kiehl, K. A., Smith, A. M., Hare, R. D., Mendrek, A., Forster, B. B., Brink, J., & Liddle, P. F. (2001). Limbic abnormalities in affective processing by criminal psychopaths as revealed by functional magnetic resonance imaging. *Biological Psychiatry, 50,* 677–684.

Kim-Cohen, J., Caspi, A., Taylor, A., Williams, B., Newcombe, R., Craig, I. W., & Moffitt, T. E. (2006). MAOA, maltreatment, and gene-environment interaction predicting children's mental health: New evidence and a meta-analysis. *Molecular Psychiatry, 11,* 903–913.

Kimbrell, T. A., George, M. S., Parekh, P. I., Ketter, R. A., Podell, D. M., Danielson, A. L., Repella, J. D., Benson, B. E., Willis, M. W., Herscovitch, P., & Post, R. M. (1999). Regional brain activity during transient self-induced anxiety and anger in healthy adults. *Biological Psychiatry, 46,* 454–465.

King, J. A., Blair, R. J. R., Mitchell, D. G. V., Dolan, R. J., & Burgess, N. (2006). Doing the right thing: A common neural circuit for appropriate violent or compassionate behavior. *NeuroImage, 30,* 1069–1076.

Knight, G. P., Fabes, R. A., & Higgins, D. A. (1996). Concerns about drawing causal inferences from meta-analyses: An example in the study of gender differences in aggression. *Psychological Bulletin, 119,* 410–421.

Knight, G. P., Guthrie, I. K., Page, M. C., & Fabes, R. A. (2002). Emotional arousal and gender differences in aggression: A meta-analysis. *Aggressive Behavior, 28,* 366–393.

Krämer, U. M., Büttner, S., Roth, G., & Münte, T. F. (in press). Trait aggressiveness modulates neurophysiological correlates of laboratory-induced reactive aggression in humans. *Journal of Cognitive Neuroscience.*

Krämer, U. M., Jansma, H., Tempelmann, C., & Münte, T. F. (2007). Tit-for-tat: The neural basis of reactive aggression. *NeuroImage, 38,* 203–211.

Kringelbach, M. L., & Rolls, E. T. (2004). The functional neuroanatomy of the human orbitofrontal cortex: Evidence from neuroimaging and neuropsychology. *Progress in Neurobiology, 72,* 341–372.

Laakso, M. P., Gunning-Dixon, F., Vaurio, O., Repo-Tiihonen, E., Soininen, H., & Tiihonen, J. (2002). Prefrontal volumes in habitually violent subjects with antisocial personality disorder and type 2 alcoholism. *Psychiatry Research: Neuroimaging, 114,* 95–102.

LaGrange, T. C., & Silverman, R. A. (1999). Low self-control and opportunity: Testing the general theory of crime as an explanation for gender differences in delinquency. *Criminology, 37,* 41–72.

Lapierre, D., Braun, C. M. J., & Hodgins, S. (1995). Ventral frontal deficits in psychopathy: Neuropsychological test findings. *Neuropsychologia, 33,* 139–151.

Lee, R., & Coccaro, E. (2001). The neuropsychopharmacology of criminality and aggression. *Canadian Journal of Psychiatry, 46,* 35–44.

Lee, J. H., Kim, H. T., & Hyun, D. S. (2003). Possible association between serotonin transporter promoter region polymorphism and impulsivity in Koreans. *Psychiatry Research, 118,* 19–24.

Lesch, K. P., & Merschdorf, U. (2000). Impulsivity, aggression, and serotonin: A molecular psychobiological perspective. *Behavioral Sciences and the Law, 18,* 581–604.

Leyton, M., Okazawa, H., Diksic, M., Joel, P., Rosa, P., Mzengeza, S., Young, S. N., Blier, P., & Benkelfat, C. (2001). Brain regional alpha-[11C]methyl-L-tryptophan trapping in impulsive subjects with borderline personality disorder. *American Journal of Psychiatry, 158,* 775–782.

Loney, B. R., Butler, M., Lima, E., Counts, C., & Eckel, L. (2006). The relation between salivary cortisol, calous-unemotional traits, and conduct problems in an adolescent non-referred sample. *Journal of Child Psychology and Psychiatry and Allied Disciplines, 47,* 30–36.

Lorber, M. F. (2004). Psychophysiology of aggression, psychopathy, and conduct problems: A meta-analysis. *Psychological Bulletin, 130,* 531–552.

Lotze, M., Veit, R., Anders, S., & Birbaumer, N. (2007). Evidence for a different role of the ventral and dorsal medial prefrontal cortex for social reactive aggression: An interactive fMRI study. *NeuroImage, 34,* 470–478.

Mah, L. W., Arnold, M. C., & Grafman, J. (2005). Deficits in social knowledge following damage to ventromedial prefrontal cortex. *Journal of Neuropsychiatry and Clinical Neurosciences, 17,* 66–74.

Mann, J. J. (2003). Neurobiology of suicidal behaviour. *Nature Reviews Neuroscience, 4,* 819–828.

Manuck, S. B., Flory, J. D., McCaffery, J. M., Matthews, K. A., Mann, J. J., & Muldoon, M. F. (1998). Aggression, impulsivity, and central nervous system serotonergic responsivity in a non-patient sample. *Neuropsychopharmacology, 19,* 287–299.

Manuck, S. B., Flory, J. D., Ferrell, R. E., Dent, K. M., Mann, J. J., & Muldoon, M. F. (1999). Aggression and anger-related traits associated with a polymorphism of the tryptophan hydroxylase gene. *Biological Psychiatry, 45,* 603–614.

Manuck, S. B., Flory, J. D., Ferrell, R. E., Mann, J. J., & Muldoon, M. F. (2000). A regulatory polymorphism of the monoamine oxidase-A gene may be associated with variability in aggression, impulsivity, and central nervous system serotonergic responsivity. *Psychiatry Research, 95,* 9–23.

Manuck, S. B., Flory, J. D., Muldoon, M. F., & Ferrell, R. E. (2002). Central nervous system serotonergic responsivity and aggressive disposition in men. *Physiology & Behavior, 77,* 705–709.

Marsh, D. M., Dougherty, D. D., Moeller, F. G., Swann, A. C., & Spiga, R. (2002). Laboratory-measured aggressive behavior of women: Acute tryptophan depletion and augmentation. *Neuropsychopharmacology, 26,* 660–671.

Mathiak, K., & Weber, R. (2006). Toward brain correlates of natural behavior: fMRI during violent video games. *Human Brain Mapping, 27,* 948–956.

McClure, E. B., Monk, C. S., Nelson, E. E., Zarahn, E., Leibenluft, E., Bilder, R. M., Charney, D. S., Ernst, M., & Pine, D. S. (2004). A developmental examination of gender differences in brain engagement during evaluation of threat. *Biological Psychiatry, 55,* 1047–1055.

Meyer-Lindenberg, A., Buckholtz, J. W., Kolachana, B., R. Hariri, A., Pezawas, L., Blasi, G., Wabnitz, A., Honea, R., Verchinski, B., Callicott, J. H., Egan, M., Mattay, V., & Weinberger, D. R. (2006). Neural mechanisms of genetic risk for impulsivity and violence in humans. *Proceedings of the National Academy of Sciences of the United States of America, 103,* 6269–6274.

Minzenberg, M. J., Grossman, R., New, A. S., Mitropoulou, V., Yehuda, R., Goodman, M., Reynolds, D. A., Silverman, J. M., Coccaro, E. F., Marcus, S., & Siever, L. J. (2006). Blunted hormone responses to ipsapirone are associated with trait impulsivity in personality disorder patients. *Neuropsychopharmacology, 31,* 197–203.

Mitchell, D. G. V., Colledge, E., Leonard, A., & Blair, R. J. R. (2002). Risky decisions and response reversal: Is there evidence of orbitofrontal cortex dysfunction in psychopathic individuals? *Neuropsychologia, 40,* 2013–2022.

Mitchell, D. G. V., Avny, S. B., & Blair, R. J. R. (2006a). Divergent patterns of aggressive and neurocognitive characteristics in acquired versus developmental psychopathy. *Neurocase, 12,* 164–178.

Mitchell, D. G. V., Fine, C., Richell, R. A., Newman, C., Lumsden, J., Blair, K. S., & Blair, R. J. R. (2006b). Instrumental learning and relearning in individuals with psychopathy and in patients with lesions involving the amygdala or orbitofrontal cortex. *Neuropsychology, 20,* 280–289.

Moffitt, T. E., Caspi, A., Dickson, N., Silva, P., & Stanton, W. (1996). Childhood-onset versus adolescent-onset antisocial conduct problems in males: Natural history from ages 3 to 18 years. *Development and Psychopathology, 8,* 399–424.

Moffitt, T. E., Brammer, G. L., Caspi, A., Fawcett, J. P., Raleigh, M., Yuwiler, A., & Silva, P. (1998). Whole blood serotonin relates to violence in an epidemiological study. *Biological Psychiatry, 43,* 446–457.

Moffitt, T. E., Caspi, A., Rutter, M., & Silva, P. A. (2001). *Sex differences in antisocial behaviour: Conduct disorder, delinquency, and violence in the dunedin longitudinal study.* Cambridge: University Press.

Møller, S. E., Mortensen, E. L., Breum, L., Alling, C., Larsen, O. G., & Bøge-Rasmussen, T. (1996). Aggression and personality: Association with amino acids and monoamine metabolites. *Psychological Medicine, 26,* 323–331.

Monk, C. S., Webb, S. J., & Nelson, C. A. (2001). Prenatal neurobiological development: Molecular mechanisms and anatomical change. *Developmental Neuropsychology, 19,* 211–236.

Müller, J. L., Sommer, M., Wagner, V., Lange, K., Taschler, H., Röder, C. H., Schuierer, G., Klein, H. E., & Hajak, G. (2003). Abnormalities in emotion processing within cortical and subcortical regions in criminal psychopaths: Evidence from a functional magnetic resonance imaging study using pictures with emotional content. *Biological Psychiatry, 54,* 152–162.

Munafò, M. R., Clark, T. G., Moore, L. R., Payne, E., Walton, R., & Flint, J. (2003). Genetic polymorphisms and personality in healthy adults: A systematic review and meta-analysis. *Molecular Psychiatry, 8,* 471–484.

Netter, P., Hennig, J., & Roed, I. S. (1996). Serotonin and dopamine as mediators of sensation seeking behavior. *Neuropsychobiology, 34,* 155–165.

New, A. S., Gelernter, J., Yovell, Y., Trestman, R. L., Nielsen, D. A., Silverman, J., Mitropoulou, V., & Siever, L. J. (1998). Tryptophan hydroxylase genotype is associated with impulsive-aggression measures: A preliminary study. *American Journal of Medical Genetics, 81,* 13–17.

New, A. S., Hazlett, E. A., Buchsbaum, M. S., Goodman, M., Reynolds, D., Mitropoulou, V., Sprung, L., Shaw, R. B., Koenigsberg, H., Platholi, J., Silverman, J., & Siever, L. (2002). Blunted prefrontal cortical [18]flurodeoxyglucose positron emission tomography response to meta-chlorophenylpiperazine in impulsive aggression. *Archives of General Psychiatry, 59,* 621–629.

New, A. S., Buchsbaum, M. S., Hazlett, E. A., Goodman, M., Koenigsberg, H. W., Lo, J., Iskander, I., Newmark, R., Brand, J., O'Flynn, K., & Siever, L. J. (2004). Fluoxetine increases relative metabolic rate in prefrontal cortex in impulsive aggression. *Psychopharmacology, 176,* 451–458.

New, A. S., Trestman, R. F., Mitropoulou, V., Goodman, M., Koenigsberg, H. H., Silverman, J., & Siever, L. J. (2004). Low prolactin response to fenfluramine in impulsive aggression. *Journal of Psychiatric Research, 38,* 223–230.

Nishizawa, S., Benkelfat, C., Young, S. N., Leyton, M., Mzengeza, S., Demontigny, C., Blier, P., & Diksic, M. (1997). Differences between males and females in rates of serotonin synthesis in human brain. *Proceedings of the National Academy of Sciences of the United States of America, 94,* 5308–5313.

Nomura, M., Ohira, H., Haneda, K., Iidaka, T., Sadato, N., Okada, T., & Yonekura, Y. (2004). Functional association of the amygdala and ventral prefrontal cortex during cognitive evaluation of facial expressions primed by masked angry faces: An event-related fMRI study. *NeuroImage, 21,* 352–363.

O'Leary, M. M., Loney, B. R., & Eckel, L. A. (2007). Gender differences in the association between psychopathic personality traits and cortisol response to induced stress. *Psychoneuroendocrinology, 32,* 183–191.

Oquendo, M. A., Placidi, G. P. A., Malone, K. M., Campbell, C., Keilp, J., Brodsky, B., Kegeles, L. S., Cooper, T. B., Parsey, R. V., Van Heertum, R. L., & Mann, J. J. (2003). Positron emission tomography of regional brain metabolic responses to a serotonergic challenge and lethality of suicide attempts in major depression. *Archives of General Psychiatry, 60,* 14–22.

Parsey, R. V., Oquendo, M. A., Simpson, N. R., Ogden, R. T., Van Heertum, R., Arango, V., & Mann, J. J. (2002). Effects of sex, age, and aggressive traits in man on brain serotonin 5-HT$_{1A}$ receptor binding potential measured by PET using [C-11]WAY-100635. *Brain Research, 954,* 173–182.

Passamonti, L., Fera, F., Magariello, A., Cerasa, A., Gioia, M. C., Muglia, M., Nicoletti, G., Gallo, O., Provinciali, L., & Quattrone, A. (2006). Monoamine oxidase-A genetic variations influence brain activity associated with inhibitory control: New insight into the neural correlates of impulsivity. *Biological Psychiatry, 59,* 334–340.

Patrick, C. J., Bradley, M. M., & Lang, P. J. (1993). Emotion in the criminal psychopath: Startle reflex modulation. *Journal of Abnormal Psychology, 102,* 82–92.

Phillips, M. L., Williams, L., Senior, C., Bullmore, E. T., Brammer, M. J., Andrew, C., Williams, S. C. R., & David, A. S. (1999). A differential neural response to threatening and non-threatening negative facial expressions in paranoid and non-paranoid schizophrenics. *Psychiatry Research: Neuroimaging, 92,* 11–31.

Phillips, M. L., Drevets, W. C., Rauch, S. L., & Lane, R. (2003a). Neurobiology of emotion perception I: The neural basis of normal emotion perception. *Biological Psychiatry, 54,* 504–514.

Phillips, M. L., Drevets, W. C., Rauch, S. L., & Lane, R. (2003b). Neurobiology of emotion perception II: Implications for major psychiatric disorders. *Biological Psychiatry, 54,* 515–528.

Pietrini, P., Guazzelli, M., Basso, G., Jaffe, K., & Grafman, J. (2000). Neural correlates of imaginal aggressive behavior assessed by positron emission tomography in healthy subjects. *American Journal of Psychiatry, 157,* 1772–1781.

Placidi, G. P. A., Oquendo, M. A., Malone, K. M., Huang, Y. Y., Ellis, S. P., & Mann, J. J. (2001). Aggressivity, suicide attempts, and depression: Relationship to cerebrospinal fluid monoamine metabolite levels. *Biological Psychiatry, 50,* 783–791.

Popma, A., Jansen, L. M. C., Vermeiren, R., Steiner, H., Raine, A., Van Goozen, S. H. M., van Engeland, H., & Doreleijers, T. A. H. (2006). Hypothalamus pituitary adrenal axis and autonomic activity during stress in delinquent male adolescents and controls. *Psychoneuroendocrinology, 31,* 948–957.

Popma, A., Vermeiren, R., Geluk, C. A. M. L., Rinne, T., van den Brink, W., Knol, D. L., Jansen, L. M. C., van Engeland, H., & Doreleijers, T. A. H. (2007). Cortisol moderates the relationship between testosterone and aggression in delinquent male adolescents. *Biological Psychiatry, 61,* 405–411.

Potegal, M., & Archer, J. (2004). Sex differences in childhood anger and aggression. *Child and Adolescent Psychiatric Clinics of North America, 13,* 513–528.

Raine, A., Venables, P. H., & Williams, M. (1990). Relationships between central and autonomic measures of arousal at age 15 years and criminality at age 24 years. *Archives of General Psychiatry, 36,* 1457–1464.

Raine, A., Buchsbaum, M. S., Stanley, J., Lottenberg, S., Abel, L., & Stoddard, J. (1994). Selective reductions in prefrontal glucose metabolism in murderers. *Biological Psychiatry, 36,* 365–373.

Raine, A., Buchsbaum, M., & Lacasse, L. (1997). Brain abnormalities in murderers indicated by positron emission tomography. *Biological Psychiatry, 42,* 495–508.

Raine, A., Venables, P. H., & Mednick, S. A. (1997). Low resting heart rate at age 3 years predisposes to aggression at age 11 years – evidence from the mauritius child health project. *Journal of the American Academy of Child & Adolescent Psychiatry, 36,* 1457–1464.

Raine, A., Meloy, J. R., Bihrle, S., Stoddard, J., Lacasse, L., & Buchsbaum, M. S. (1998). Reduced prefrontal and increased subcortical brain functioning assessed using positron emission tomography in predatory and affective murderers. *Behavioral Sciences and the Law, 16,* 319–332.

Raine, A., Lencz, T., Bihrle, S., LaCasse, L., & Colletti, P. (2000). Reduced prefrontal gray matter volume and reduced autonomic activity in antisocial personality disorder. *Archives of General Psychiatry, 57,* 119–127.

Ramirez, J. M., & Andreu, J. M. (2006). Aggression, and some related psychological constructs (anger, hostility, and impulsivity) Some comments from a research project. *Neuroscience and Biobehavioral Reviews, 30,* 276–291.

Reif, A., & Lesch, K. P. (2003). Toward a molecular architecture of personality. *Behavioural Brain Research, 139,* 1–20.

Retz, W., Retz-Junginger, P., Supprian, T., Thome, J., & Rösler, M. (2004). Association of serotonin transporter promoter gene polymorphism with violence: Relation with personality disorders, impulsivity, and childhood ADHD psychopathology. *Behavioral Sciences and the Law, 22,* 415–425.

Roberti, J. W. (2004). A review of behavioral and biological correlates of sensation seeking. *Journal of Research in Personality, 38,* 256–279.

Rolls, E. T., Hornak, J., Wade, D., & McGrath, C. (1994). Emotion-related learning in patients with social and emotional changes associated with frontal lobe damage. *Journal of Neurology, Neurosurgery and Psychiatry, 57,* 1518–1524.

Rotondo, A., Schuebel, K. E., Bergen, A. W., Aragon, R., Virkkunen, M., Linnoila, M., Goldman, D., & Nielsen, D. A. (1999). Identification of four variants in the tryptophan hydroxylase promoter and association to behavior. *Molecular Psychiatry, 4,* 360–368.

Rubia, K., Lee, F., Cleare, A. J., Tunstall, N., Fu, C. H. Y., Brammer, M., & McGuire, P. (2005). Tryptophan depletion reduces right inferior prefrontal activation during response inhibition in fast, event-related fMRI. *Psychopharmacology, 179,* 791–803.

Rubinow, D. R., Schmidt, P. J., & Roca, C. A. (1998). Estrogen-serotonin interactions: Implications for affective regulation. *Biological Psychiatry, 44,* 839–850.

Rujescu, D., Giegling, I., Sato, T., Hartmann, A. M., & Möller, H. J. (2003). Genetic variation in tryptophan hydroxylase in suicidal behavior: Analysis and meta-analysis. *Biological Psychiatry, 54,* 465–473.

Schneider, F., Habel, U., Kessler, C., Posse, S., Grodd, W., & Muller-Gartner, H. W. (2000). Functional imaging of conditioned aversive emotional responses in antisocial personality disorder. *Neuropsychobiology, 42,* 192–201.

Shirtcliff, E. A., Granger, D. A., Booth, A., & Johnson, D. (2005). Low salivary cortisol levels and externalizing behavior problems in youth. *Development and Psychopathology, 17,* 167–184.

Siever, L. J. (2005). Endophenotypes in the personality disorders. *Dialogues in Clinical Neuroscience, 7,* 139–151.

Siever, L. J., Buchsbaum, M. S., New, A. S., Spiegel-Cohen, J., Wei, T., Hazlett, E. A., Sevin, E., Nunn, M., & Mitropoulou, V. (1999). D,L-Fenfluramine response in impulsive personality disorder assessed with [^{18}F]fluorodeoxyglucose positron emission tomography. *Neuropsychopharmacology, 20,* 413–423.

Soderstrom, H., Hultin, L., Tullberg, M., Wikkelso, C., Ekholm, S., & Forsman, A. (2002). Reduced fronto-temporal perfusion in psychopathic personality. *Psychiatry Research: Neuroimaging, 114,* 81–94.

Soloff, P. H., Lynch, K. G., & Moss, H. B. (2000a). Serotonin, impulsivity, and alcohol use disorders in the older adolescent: A psychobiological study. *Alcoholism: Clinical and Experimental Research, 24,* 1609–1619.

Soloff, P. H., Meltzer, C. C., Greer, P. J., Constantine, D., & Kelly, T. M. (2000b). A fenfluramine-activated FDG-PET study of borderline personality disorder. *Biological Psychiatry, 47,* 540–547.

Soloff, P. H., Kelly, T. M., Strotmeyer, S. J., Malone, K. M., & Mann, J. J. (2003). Impulsivity, gender, and response to fenfluramine challenge in borderline personality disorder. *Psychiatry Research, 119,* 11–24.

Soloff, P. H., Meltzer, C. C., Becker, C., Greer, P. J., Kelly, T. M., & Constantine, D. (2003). Impulsivity and prefrontal hypometabolism in borderline personality disorder. *Psychiatry Research: Neuroimaging, 123,* 153–163.

Sprengelmeyer, R., Rausch, M., Eysel, U. T., & Przuntek, H. (1998). Neural structures associated with recognition of facial expressions of basic emotions. *Proceedings of the Royal Society of London. Series B: Biological Sciences, 265,* 1927–1931.

Stalenheim, E. G., Eriksson, E., von Knorring, L., & Wide, L. (1998). Testosterone as a biological marker in psychopathy and alcoholism. *Psychiatry Research, 77,* 79–88.

Staner, L., Uyanik, G., Correa, H., Tremeau, F., Monreal, J., Crocq, M.-A., Stefos, G., Morris-Rosendahl, D. J., & Macher, J. P. (2002). A dimensional impulsive-aggressive phenotype is associated with the A218C polymorphism of the tryptophan hydroxylase gene: A pilot study in well-characterized impulsive inpatients. *American Journal of Medical Genetics, 114,* 553–557.

Stanley, B., Molcho, A., Stanley, M., Winchel, R., Gameroff, M. J., Parsons, B., & Mann, J. J. (2000). Association of aggressive behavior with altered serotonergic function in patients who are not suicidal. *American Journal of Psychiatry, 157,* 609–614.

Sterzer, P., Stadler, C., Krebs, A., Kleinschmidt, A., & Poustka, F. (2005). Abnormal neural responses to emotional visual stimuli in adolescents with conduct disorder. *Biological Psychiatry, 57,* 7–15.

Sterzer, P., Stadler, C., Poustka, F., & Kleinschmidt, A. (2007). A structural neural deficit in adolescents with conduct disorder and its association with lack of empathy. *NeuroImage, 37,* 335–342.

Strauss, M. M., Makris, N., Aharon, I., Vangel, M. G., Goodman, J., Kennedy, D. N., Gasic, G. P., & Breiter, H. C. (2005). fMRI of sensitization to angry faces. *NeuroImage, 26,* 389–413.

Tarullo, A. R., & Gunnar, M. R. (2006). Child maltreatment and the developing HPA axis. *Hormones and Behavior, 50,* 632–639.

Tebartz van Elst, L., Hesslinger, B., Thiel, T., Geiger, E., Haegele, K., Lemieux, L., Lieb, K., Bohus, M., Hennig, J., & Ebert, D. (2003). Frontolimbic brain abnormalities in patients with borderline personality disorder: A volumetric magnetic resonance imaging study. *Biological Psychiatry, 54,* 163–171.

Teicher, M. H., Andersen, S. L., Polcari, A., Anderson, C. M., Navalta, C. P., & Kim, D. M. (2003). The neurobiological consequences of early stress and childhood maltreatment. *Neuroscience and Biobehavioral Reviews, 27,* 33–44.

Tremblay, R. E., & Nagin, D. S. (2005). The developmental origins of physical aggression in humans. In R. E. Tremblay, W. W. Hartup & J. Archer (Eds.), *Developmental Origins of Aggression* (pp. 83–106). New York: Guilford.

van Goozen, S. H. M., & Fairchild, G. (2006). Neuroendocrine and neurotransmitter correlates in children with antisocial behavior. *Hormones and Behavior, 50,* 647–654.

van Honk, J., Tuiten, A., Verbaten, R., van den Hout, M., Koppeschaar, H., Thijssen, J., & de Haan, E. (1999). Correlations among salivary testosterone, mood, and selective attention to threat in humans. *Hormones and Behavior, 36,* 17–24.

van Honk, J., Hermans, E. J., Putman, P., Montagne, B., & Schutter, D. J. L. G. (2002). Defective somatic markers in sub-clinical psychopathy. *Neuroreport, 13,* 1025–1027.

van Honk, J., Schutter, D. J. L. G., Hermans, E. J., Putman, P., Tuiten, A., & Koppeschaar, H. (2004). Testosterone shifts the balance between sensitivity for punishment and reward in healthy young women. *Psychoneuroendocrinology, 29,* 937–943.

Veit, R., Flor, H., Erb, M., Hermann, C., Lotze, M., Grodd, W., & Birbaumer, N. (2002). Brain circuits involved in emotional learning in antisocial behavior and social phobia in humans. *Neuroscience Letters, 328,* 233–236.

Verona, E., & Kilmer, A. (2007). Stress exposure and affective modulation of aggressive behavior in men and women. *Journal of Abnormal Psychology, 116,* 410–421.

Verona, E., Joiner, T. E., Johnson, F., & Bender, T. W. (2006). Gender specific gene-environment interactions on laboratory-assessed aggression. *Biological Psychology, 71,* 33–41.

Viding, E., Blair, R. J. R., Moffitt, T. E., & Plomin, R. (2005). Evidence for substantial genetic risk for psychopathy in 7-year-olds. *Journal of Child Psychology and Psychiatry, 46,* 592–597.

Virkkunen, M., Kallio, E., Rawlings, R., Tokola, R., Poland, R. E., Guidotti, A., Nemeroff, C., Bissette, G., Kalogeras, K., Karonen, S. L., et al. (1994). Personality profiles and state aggressiveness in Finnish alcoholic, violent offenders, fire setters, and healthy volunteers. *Archives of General Psychiatry, 51,* 28–33.

von der Pahlen, B., Lindman, R., Sarkola, T., Makisalo, H., & Eriksson, C. J. P. (2002). An exploratory study on self-evaluated aggression and androgens in women. *Aggressive Behavior, 28,* 273–280.

Watts-English, T., Fortson, B. L., Gibler, N., Hooper, S. R., & De Bellis, M. D. (2006). The psychobiology of maltreatment in childhood. *Journal of Social Issues, 62,* 717–736.

Webb, S. J., Monk, C. S., & Nelson, C. A. (2001). Mechanisms of postnatal neurobiological development: Implications for human development. *Developmental Neuropsychology, 19,* 147–171.

Whalen, P. J., Shin, L. M., McInerney, S. C., Fischer, H., Wright, C. I., & Rauch, S. L. (2001). A functional MRI study of human amygdala responses to facial expressions of fear versus anger. *Emotion, 1,* 70–83.

Widom, C. S., & Brzustowicz, L. M. (2006). MAOA and the 'cycle of violence': Childhood abuse and neglect, MAOA genotype, and risk for violent and antisocial behavior. *Biological Psychiatry, 60,* 684–689.

Wihlbäck, A.-C., Sunderström, P., Bixo, M., Allard, P., Mjörndal, T., & Spigset, O. (2004). Influence of menstrual cycle on platelet serotonin uptake site and serotonin$_{2A}$ receptor binding. *Psychoneuroendocrinology, 29,* 757–766.

Williams, R. B., Marchuk, D. A., Gadde, K. M., Barefoot, J. C., Grichnik, K., Helms, M. J., Kuhn, C. M., Lewis, J. G., Schanberg, S. M., Stafford-Smith, M., Suarez, E. C., Clary, G. L., Svenson, I. K., & Siegler, I. C. (2003). Serotonin-related gene polymorphisms and central nervous system serotonin function. *Neuropsychopharmacology, 28,* 533–541.

Wingrove, J., Bond, A., Cleare, A., & Sherwood, R. (1999a). Plasma tryptophan and trait aggression. *Psychopharmacology, 13,* 235–237.

Wingrove, J., Bond, A. J., Cleare, A. J., & Sherwood, R. (1999b). Trait hostility and prolactin response to tryptophan enhancement/depletion. *Neuropsychobiology, 40,* 202–206.

Woermann, F. G., Tebartz van Elst, L. T., Koepp, M. J., Free, S. L., Thompson, P. J., Trimble, M. R., & Duncan, J. S. (2000). Reduction of frontal neocortical grey matter associated with affective aggression in patients with temporal lobe epilepsy: An objective voxel by voxel analysis of automatically segmented MRI. *Journal of Neurology, Neurosurgery & Psychiatry, 68,* 162–169.

Yang, Y., Raine, A., Lencz, T., Bihrle, S., LaCasse, L., & Colletti, P. (2005). Volume reduction in prefrontal gray matter in unsuccessful criminal psychopaths. *Biological Psychiatry, 57,* 1103–1108.

Zahn, T. P., Grafman, J., & Tranel, D. (1999). Frontal lobe lesions and electrodermal activity: Effects of significance. *Neuropsychologia, 37,* 1227–1241.

Zalsman, G., Frisch, A., Bromberg, M., Gelernter, J., Michaelovsky, E., Campino, A., Erlich, Z., Tyano, S., Apter, A., & Weizman, A. (2001). Family-based association study of serotonin transporter promoter in suicidal adolescents: No association with suicidality but possible role in violence traits. *American Journal of Medical Genetics, 105,* 239–245.

Zill, P., Büttner, A., Eisenmenger, W., Möller, H. J., Bondy, B., & Ackenheil, M. (2004). Single nucleotide polymorphism and haplotype analysis of a novel tryptophan hydroxylase isoform (TPH2) gene in suicide victims. *Biological Psychiatry, 56,* 581–586.

NEUROCASE

Group and single case investigations of brain–behavior relationships in adults and children

Abstracted/Indexed in: MEDLINE, SciSearch, Research Alert, Neuroscience Citation Index, Chemical Abstracts, EMBASE and CINAHL database, Cumulative Index to Nursing & Allied Health Literature print index, and PSYINFO.

Neurocase (ISSN: 1335-4794) is published bi-monthly (in February, April, June, August, October, and December) for a total of 6 issues per year by Psychology Press in the UK, an imprint of the Taylor & Francis Group, an Informa business. Distributed in the US by Taylor & Francis, 325 Chestnut St – 8th Fl., Philadelphia, PA 19106, USA. Periodicals Postage is paid at Philadelphia, PA and additional mailing offices. **US Postmaster:** Please send address changes to *Neurocase*, Taylor & Francis, 325 Chestnut St – 8th Fl., Philadelphia, PA 19106, USA.

Subscription orders and changes of address should be addressed to: Psychology Press, T & F Customer Services, Informa UK Ltd., Sheepen Place, Colchester, Essex, CO3 3LP, UK. E-mail: tf.enquiries@tfinforma.com; Tel: 020 7017 5544; Fax 020 7017 5198.

Annual Subscription, Volume 14, 2008
Print ISSN - 1355-4794.

An institutional subscription to the print edition includes free access to the online edition for any number of concurrent users across a local area network.

Institutions (full sub'n):	£535.00 (UK);	€705.00 (Europe);	$881.00 (RoW).
Institutions (online only):	£508.00 (UK);	€669.00 (Europe);	$836.00 (RoW).

Subscriptions purchased at the personal (print only) rate are strictly for personal, non-commercial use. The reselling of personal subscriptions is prohibited. Personal subscriptions must be purchased with a personal cheque or credit card. Proof of personal status may be requested.

Individuals:	£237.00 (UK);	€314.00 (Europe);	$392.00 (RoW).

Neurocase now offers an iOpenAccess option for authors. For more information, see: **www.tandf.co.uk/journals/iopenaccess.asp**

Production and Advertising Office: Psychology Press, 2 Park Square, Milton Park, Abingdon, Oxon, OX14 4RN

Neurocase is available online. Please go to http://www.psypress.com/journals.asp for current information about this journal including how to access the online version or register for the free table of contents alerting service. Information about Psychology Press journals and other publications is also available at http://www.psypress.com/journals.asp.

Instructions for Authors

Aims and Scope

Neurocase is a rapid response journal of both adult and child case studies in neuropsychology, neuropsychiatry and behavioral neurology. Four types of manuscripts are considered for publication: single case investigations that bear directly on issues of relevance to theoretical issues or brain–behavior relationships; group studies of subjects with brain dysfunction that address issues relevant to the understanding of human cognition; reviews of important topics in the domains of neuropsychology, neuropsychiatry, and behavioral neurology; and brief reports (up to 2,500 words) that replicate previous reports dealing with issues of considerable significance. Investigations of the anatomic basis of behavior and cognition by means of static or functional brain imaging are encouraged. Topic reviews are included in most issues.

Manuscripts should be submitted to **http://mc.manuscriptcentral.com/nncs**

Each manuscript must be accompanied by a statement that it has not been published elsewhere and that it has not been submitted simultaneously for publication elsewhere. Authors are responsible for obtaining permission to reproduce copyrighted material from other sources and are required to sign an agreement for the transfer of copyright to the publisher. All accepted manuscripts, artwork, and photographs become the property of the publisher.

All parts of the manuscript should be typewritten, double-spaced, with margins of at least one inch on all sides. Number manuscript pages consecutively throughout the paper. Authors should also supply up to 5 Keywords, and a shortened version of the title suitable for the running head, not exceeding 50 character spaces. Each article should be summarized in an abstract of not more than 100 words. Avoid abbreviations, diagrams, and reference to the text in the abstract.

References

Cite in the text by author and date (Tyler & Moss, 2001). Prepare reference list in accordance with the APA Style Manual (5th ed.). Examples:

Journal: Tyler, L. K., & Moss, H. E. (2001). Towards a distributed account of conceptual knowledge. *Trends in Cognitive Sciences, 5,* 244–252.

Book: Moore, D. S., & McCabe, G. P. (1989). *Introduction to the practice of statistics.* New York: W.H. Freeman and Company.

Book chapter: Sartori, G., Miozzo, M., & Job, R. (1994). Rehabilitation of semantic memory impairments. In M. J. Riddoch & G. W. Humphreys (Eds.), *Cognitive neuropsychology and cognitive rehabilitation* (pp. 103–124). Hove, UK: Lawrence Erlbaum Associates Ltd.

Illustrations

Illustrations submitted (line drawings, halftones, photos, photomicrographs, etc.) should be clean originals or digital files. Digital files are recommended for highest quality reproduction and should follow these guidelines:

- 300 dpi or higher
- Sized to fit on journal page
- EPS, TIFF, or PSD format only
- Submitted as separate files, not embedded in text files

Color illustrations will be considered for publication; however, the author will be required to bear the full cost involved in their printing and publication. The charge for the first page with color is $900.00. The next three pages with color are $450.00 each. A custom quote will be provided for color art totaling more than 4 journal pages. Good-quality color prints should be provided in their final size. The publisher has the right to refuse publication of color prints deemed unacceptable.

Tables and Figures

Tables and figures (illustrations) should not be embedded in the text, but should be included as separate sheets or files. A short descriptive title should appear above each table with a clear legend and any footnotes suitably identified below. All units must be included. Figures should be completely labeled, taking into account necessary size reduction. Captions should be typed, double-spaced, on a separate sheet. All original figures should be clearly marked in pencil on the reverse side with the number, author's name, and top edge indicated.

Proofs

Page proofs are sent to the designated author using Taylor & Francis' EProof system. They must be carefully checked and returned within 48 hours of receipt.

Reprints

Each corresponding author will receive one copy of the issue in which the article appears. Reprints of individual articles are available for order at the time authors review page proofs. A discount on reprints is available to authors who order before print publication.

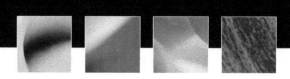

AUTHOR SERVICES

Publish With Us

 Taylor & Francis
Taylor & Francis Group

 Routledge
Taylor & Francis Group

 Psychology Press
Taylor & Francis Group

informa
healthcare

The Taylor & Francis Group Author Services Department aims to enhance your publishing experience as a journal author and optimize the impact of your article in the global research community. Assistance and support is available, from preparing the submission of your article through to setting up citation alerts post-publication on **informa**world™, our online platform offering cross-searchable access to journal, book and database content.

Our Author Services Department can provide advice on how to:

- direct your submission to the correct journal
- prepare your manuscript according to the journal's requirements
- maximize your article's citations
- submit supplementary data for online publication
- submit your article online via Manuscript Central™
- apply for permission to reproduce images
- prepare your illustrations for print
- track the status of your manuscript through the production process
- return your corrections online
- purchase reprints through Rightslink™
- register for article citation alerts
- take advantage of our i*OpenAccess* option
- access your article online
- benefit from rapid online publication via i*First*

See further information at:
www.informaworld.com/authors

or contact:
Author Services Manager, Taylor & Francis, 4 Park Square, Milton Park, Abingdon, Oxon OX14 4RN, UK, email: authorqueries@tandf.co.uk

Routledge
Taylor & Francis Group

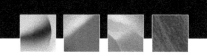

Philosophical Psychology

EDITORS:

William Bechtel, *University of California, San Diego, USA*
Cees van Leeuwen, *Laboratory for Perceptual Dynamics, Brain Science Institute, RIKEN, Saitama, Japan*

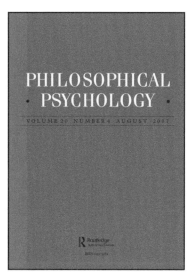

Philosophical Psychology is an international journal devoted to developing and strengthening the links between philosophy and the psychological sciences, both as basic sciences and as employed in applied settings, by publishing original, peer-refereed contributions to this expanding field of study and research.

Published articles deal with issues that arise in the cognitive and brain sciences, and in areas of applied psychology. Emphasis is placed on articles concerned with cognitive and perceptual processes, models of psychological processing, including neural network and dynamical systems models, and relations between psychological theories and accounts of neural underpinnings or environmental context. The journal also publishes theoretical articles concerned with the nature and history of psychology, the philosophy of science as applied to psychology, and explorations of the underlying issues - theoretical and ethical - in contemporary educational, clinical, occupational and health psychology.

Philosophical Psychology is included in the Psychology, Multidisciplinary and Ethics categories of the © Thompson ISI Journal Citation Reports 2006

SUBSCRIPTION RATES

2008 - Volume 21 (6 issues per year)
Print ISSN 0951-5089
Online ISSN 1465-394X
Institutional rate (print and online): US$1395; £845; €1116
Institutional rate (online access only): US$1325; £802; €1060
Personal rate (print only): US$333; £200; €266

T - #0943 - 101024 - C10 - 272/202/6 - PB - 9781841698540 - Gloss Lamination